A GUIDEBOOK TO BIOCHEMISTRY

FOURTH EDITION

A Guidebook to Biochemistry

FOURTH EDITION (1980)

MICHAEL YUDKIN
Tutorial Fellow of University College, Oxford, and
University Lecturer in Biochemistry

ROBIN OFFORD
Professor of Medical Biochemistry, University of Geneva;
formerly University Lecturer, and Tutor in Biochemistry
at Christ Church, Oxford

A new edition of 'A Guidebook to Biochemistry' by
K. HARRISON

CAMBRIDGE UNIVERSITY PRESS
Cambridge
London   New York   New Rochelle
Melbourne   Sydney

Published by the Press Syndicate of the University of Cambridge
The Pitt Building, Trumpington Street, Cambridge CB2 1RP
32 East 57th Street, New York, NY 10022, USA
296 Beaconsfield Parade, Middle Park, Melbourne 3206, Australia

Fourth edition of *A Guidebook to Biochemistry* by K. Harrison
© Cambridge University Press 1965

Completely revised version by Yudkin and Offord first published 1971
Reprinted 1973, 1978
Persian translation 1971
German translation 1972
Japanese translation 1975
Malay translation 1980
Fourth edition 1980

Printed in Great Britain by J. W. Arrowsmith Ltd, Bristol

British Library Cataloguing in Publication Data

Yudkin, Michael
A guidebook to biochemistry.—4th ed.
1. Biological chemistry
I. Title   II. Offord, Robin
574.1'92   QH345   79-41606

ISBN 0 521 23084 5 hard covers
ISBN 0 521 29794 X paperback
(1971 revised version
ISBN 0 521 08195 5 hard covers
ISBN 0 521 09654 5 paperback)

Contents

Section IV. Compartmentation and regulation

Preface to the fourth edition

The ten years that have passed since we wrote the last edition of *A Guidebook to Biochemistry* have seen as important an advance in the subject as any previous decade. Although in some areas of biochemistry there has been only filling-in of detail or minor correction of error, in others there have been discoveries that have totally overturned our previous ideas.

For the authors of a Guidebook the problems that result from these changes are more severe than they would be for authors of a large, comprehensive text. We have had to assess the significance of the new results and decide what places they should find within an introductory book. We have tried to avoid the dangers on the one hand of inappropriate conservatism and on the other of succumbing to fashion. Although we have made major textual changes in this edition we have kept the structure of the book much as in the 1971 edition – although we have introduced a whole new chapter on cell biology and produced a section to cover the important topic of regulation. No doubt some readers will be sorry that we have not allotted more space to this topic or to that; to them we express our regrets but can only say that, granted the size of the book, we have done our best to achieve a reasonable balance.

It is worth stressing that biochemistry is an experimental subject, and that all the facts and ideas that we present here are derived, in the end, from laboratory work. Within the limitations of space of a Guidebook, the description of experiments would have been out of place. There are many larger and fuller textbooks of biochemistry, among them Lehninger's *Biochemistry*, Stryer's *Biochemistry* and our own *Comprehensible Biochemistry*.

We were very much encouraged by the kind reception given to the last edition of this book. We hope that this new edition may win a similar acceptance.

March 1980   M. D. Y.
                                                    R. E. O.

Acknowledgements

Figs. 3.6 and 5.5.1 were specially drawn for this book by an ARGUS computer, programmed and operated by Professor A. C. T. North, to whom the authors are most grateful. Figs. 4.4 and 6.1 were adapted from *The structure and action of proteins* by R. E. Dickerson and I. Geis (Harper and Row). The angle of perspective used in Fig. 3.4 was taken from that worked out by these authors for figures in their book, as were the representations of the amino-acid side chains in Table 3.1 (although the arrangement of the table is our own). Fig. 3.3 is from *The nature of the chemical bond* by L. Pauling (Cornell University Press). Fig. 6.2 is adapted from *The biochemistry of the nucleic acids* by J. N. Davidson (Methuen) and from M. F. H. Wilkins *et al.* (1955), *Nature*, **175**, 834. Fig. 6.8 is an adaptation by J. N. Davidson (*loc. cit.*) of a drawing by D. L. D. Caspar. Fig. 6.17 was inspired by the drawing by Bunji Tagawa on p. 53 of *Cell membranes* (G. Weissman, ed., HP Publishing Co., New York) although it is not a copy of it.

We thank authors and publishers for the use of copyright material noted above, and Messrs Longman for their agreement to our use of Fig. 6.18 and a few items of text from our *Comprehensible biochemistry* (in the USA: *Biochemistry*, published by Houghton Mifflin Co.).

We are very grateful to all those who have written to us with comments and corrections to the third edition: Dr R. N. Campagne and Dr H. B. F. Dixon in particular were kind enough to send us lists covering the whole text of the book.

We would like to acknowledge the exceptional care and skill shown by Mr Steven Gardiner and his colleagues of the Cambridge University Press in seeing the new edition through all its stages of production.

Conventions and abbreviations

Biochemical reactions commonly take place at or about pH 7. This poses the problem of how one should write the structural formulae of compounds which take part in acid–base equilibria. Such compounds will be wholly dissociated, partly dissociated or undissociated at pH 7 depending on the pK of their particular dissociation equilibrium.

To avoid confusion we have tried wherever possible to write the structures of the molecules we discuss in the un-ionized form, irrespective of their true pK values and their actual state of ionization at pH 7. Thus we write in equation (11), p. 130, of the synthesis of lactic acid when in fact what is produced is mainly lactate ion,

$$CH_3CHOHCOO^-,$$

balanced by a solvated proton, H_3O^+. We depart from this rule in the few cases where it would hinder rather than help comprehension (e.g. p. 101).

In the case of the hydrogen carriers NAD and NADP we ignore ionization completely (see pp. 26 and 108) and write NAD and NADP for the oxidized form and $NADH_2$ and $NADPH_2$ for the reduced form. These abbreviations are still allowed by international convention, although the symbols NAD^+, $NADP^+$, NADH and NADPH are now preferred. We have rejected this convention because we believe that it makes the events in, for example, Chapter 8 more difficult to follow.

We have followed the frequently adopted practice of neglecting the ionization of the phosphate group by writing Ⓟ for phosphate in an organic compound and P_i for the inorganic phosphate ion. Similarly, organic pyrophosphate is written Ⓟ–Ⓟ and the inorganic ion PP_i. Thus the reaction

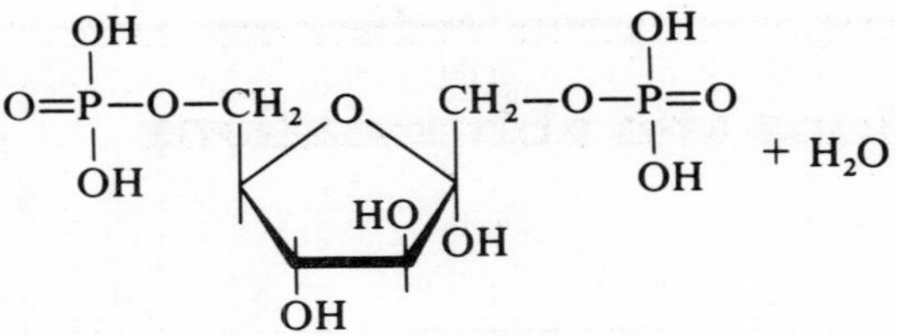

fructose-1,6-bisphosphate

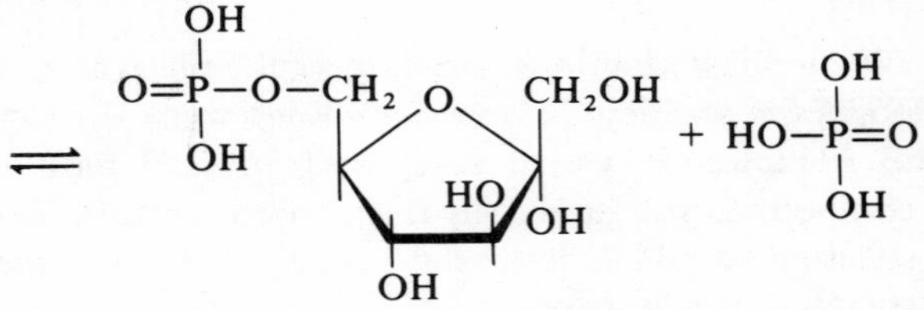

fructose-6-phosphate

(in which the possible ionizations of the phosphate groups are already neglected) is written on p. 163 as

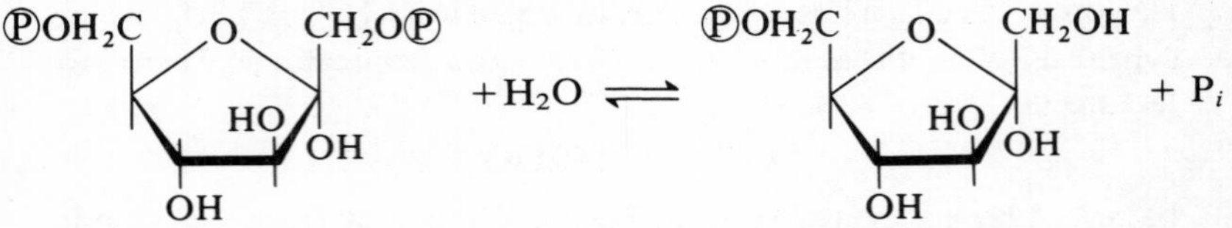

Apart from the increased rapidity with which reactions can be written, this convention has the advantage that we need no longer write H_3PO_4 as a substrate or a product and give the impression that so many biochemical reactions use or generate a strong acid.

As a consequence of neglecting ionization equilibria, equations do not always balance as regards H and OH groups. Thus if we were not to neglect ionization the equation on p. 139 should be

$$CH_3COCOO^- + CO_2 + ATP \rightleftharpoons \underset{\underset{CH_2COO^-}{|}}{COCOO^-} + ADP + P_i$$

pyruvate ion   oxaloacetate ion

rather than

$$CH_3COCOOH + CO_2 + ATP \rightleftharpoons \underset{\underset{CH_2COOH}{|}}{COCOOH} + ADP + P_i$$

as written there. It will be seen that in the latter form there is a proton missing from the left-hand side. We do not put it in the equation in case we should give the impression that it takes part in the mechanism of the main reaction.

Reversible reactions are written $\rightleftharpoons$ in the usual way.

Where an overall reaction scheme involves *coupled* reactions (see p. 99) we have sometimes found it convenient to use the Baldwin notation

This notation is not meant to imply that the reaction scheme is mechanistically irreversible. It would be possible, but clumsy, to write the arrows for the return reaction.

1 Introduction

The principal difficulty in writing a book about biochemistry is in deciding the order of the chapters. In a linear subject like mathematics, the study of more advanced concepts depends on an understanding of more elementary concepts; there is a sequence in which one progresses, and books on mathematics adhere to the sequence. Biochemistry, by contrast, is a circular subject, in which the study of any one aspect can illuminate many other aspects; it seems to us that the authors of a book on biochemistry ought to make every effort to minimize the problems that arise from this characteristic of the subject.

We have tried to overcome the difficulties in several ways. In the first place the very idea of a guidebook is that it is intended to introduce the reader to just a few outstanding features of the field. This book is not intended to be comprehensive. We have omitted many topics that a textbook (even an elementary textbook) would normally include – to cite just a few examples, we have omitted any discussion of experimental methods, any account of enzyme kinetics and all mention of some important synthetic reactions, e.g. the synthesis of urea. Our intention has been to introduce, exemplify and discuss certain crucial biochemical concepts. We have selected the topics that are best suited to the treatment of these even at the expense of seeming arbitrary in our choice of material.

The second way in which we have tried to deal with the circular nature of the subject is by introducing key ideas at certain critical points in the book. Chapter 2, which introduces an account of the macromolecular constituents of living matter, is about the forces involved in maintaining the structure of macromolecules. Chapter 7, which introduces an account of intermediary metabolism, is about the flow of free energy in metabolic reactions. Chapter 18, which introduces an account of the synthesis of informational macromolecules and regulation, is about the genetic material (DNA) and its function in determining how proteins are synthesized.

1

Thus we begin and end the book with discussions of macromolecules, which are the characteristic compounds of living organisms; and this is an example of the circularity of the subject with which we are dealing. A further means that we have used in projecting this circular subject on to a book is to include extensive cross-references; we refer not only back but also forwards. Our final attempt to help the reader to break into the subject is connected with these cross-references; it is to express the hope that, as soon as you have finished the book, you will go back to the beginning and read it again.

Section I
Structure and function
of macromolecules

2 Introduction to macromolecules

Living matter is distinguished by its reliance on the special properties of certain classes of extremely large molecule. Of these classes
the proteins, the nucleic acids and the polysaccharides are particularly prominent. These three types of molecules share certain
principles of construction, although these are at first sight obscured
by the differences which exist as a necessary consequence of the
great diversity of the functions the molecules undertake.

The principal common feature is that all three are chain polymers
formed by condensation, that is the combination of smaller molecules with the exclusion of water. In each case the smaller molecules
are drawn from a homologous series. Proteins are composed of
amino acids, of which the general formula is

$$RCHNH_2COOH$$

(see Chapter 3), polymerized by condensation between their amino
and carboxyl groups. The resulting bond between the amino acids
(which are now called amino-acid residues) is known as the peptide
bond (Fig. 2.1). The end of the chain bearing the free amino group
is called the amino terminus, that bearing the carboxyl group is
called the carboxyl terminus.

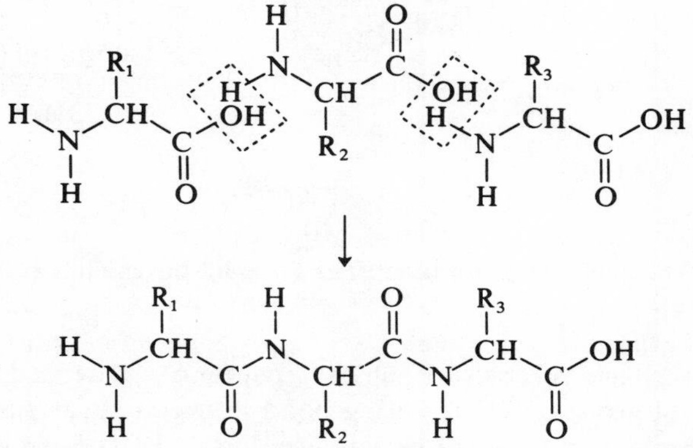

Fig. 2.1. The peptide bond. The amino terminus is at the left.

5

Similarly nucleic acids consist of nucleotides. These are of general formula

(purine or pyrimidine base)–pentose–phosphate

(see Chapter 6) and they are joined by condensation between the phosphate group of one nucleotide and an –OH group on the pentose of an adjacent nucleotide (Fig. 2.2). The linkage between the nucleotide residues is called a phosphodiester bond. One end of the chain has a pentose in which the 3′ position takes no part in the bonding and the other has a pentose in which the 5′ position takes no part. These ends are called the 3′ and 5′ ends respectively.

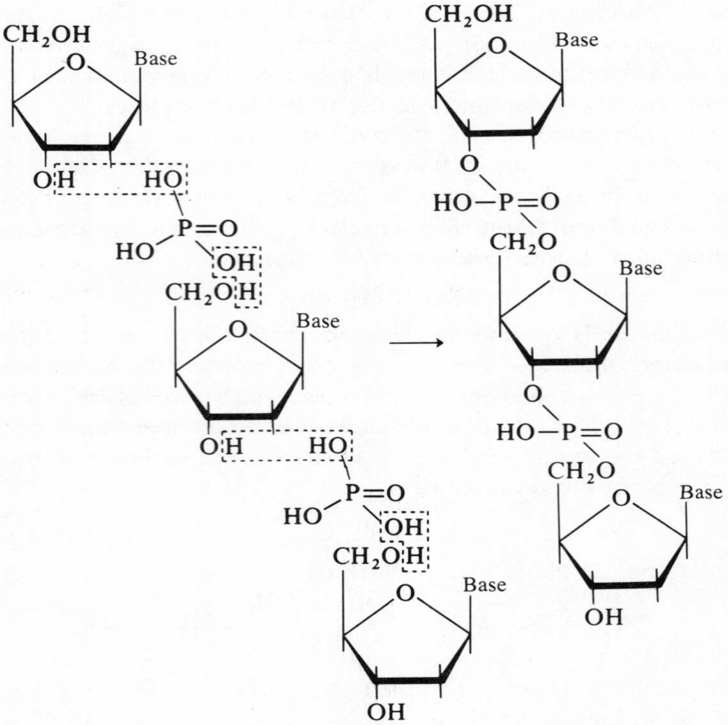

Fig. 2.2. The phosphodiester bond. The 5′ end of the chain is at the top of the Figure.

Polysaccharides consist of sugars (Chapter 6) condensed through their –OH groups. The resulting bond between sugar residues is called the glycosidic bond or linkage (Fig. 2.3). It is an essential feature of the glycosidic linkage that the C-1 –OH of one of the

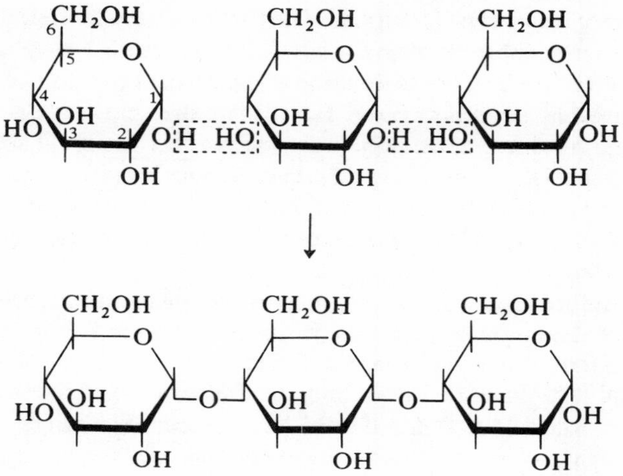

Fig. 2.3. The glycosidic bond. The reducing end is at the right.

sugars is involved. The other –OH group may belong to any of the carbon atoms of the second sugar. Because of the need for a C-1 –OH in every glycosidic bond only one end of the chain will have a sugar with position 1 free. The other end of the chain will always have a sugar with position 1 linked to the adjacent sugar. Because of the reducing properties of the uncombined 1 position these ends are called the reducing and non-reducing ends respectively.

It will be noticed that in each of these examples an unbranched polymer has been formed. However opportunities for cross-linking exist (in theory at least) in each case. In the proteins the amino acids may possess groups in the side chains which can undergo cross-linking reactions. In fact this occurs in only a few of the possible cases (see p. 33), and in normal proteins never by formation of a peptide-type bond. At points at which it does occur there can clearly be branching or joining of chains. In the nucleic acids the pentose may possess more than the two –OH groups shown and the theoretical possibility exists of branched or joined chains. They seem never to occur, however. Polysaccharides on the other hand are quite often branched. We shall see in Chapter 6 that bonds other than the glycosidic link are occasionally used for cross-linking.

It is when one looks at the nature of the homologous series involved, and the way in which selections are made from them to build up the molecule, that major differences between the macromolecules start to appear. The proteins have at their disposal about twenty different amino acids (Table 3.1) with a wide range of types of side chain. The difference when we turn to a consideration of the nucleic acids is striking. Here, the entire molecule is usually made up by drawing on only four types of nucleotide (Table 6.1). Certain specialized nucleic acids do exist which use a much greater variety of nucleotides (see p. 60), but these nucleotides are in most cases derived from the more common ones by simple chemical substitution.

The difference between the proteins and nucleic acids reflects the different demands made on them. As we shall see below, proteins have to carry out a wide range of functions, from the mechanical to the catalytic. Since the same amino acids are involved in every case, it is necessary that there should be a reasonable range, so that sufficient structural permutations are possible. Nucleic acids on the other hand have only one main type of function, the storage and transfer of information (see pp. 52, 183). Here the permutation of just four types of smaller unit suffices.

The position with the polysaccharides similarly depends on their functions. They are mostly used as structural material and for food storage (Chapter 6), and these functions do not call for such great subtlety in combination of units. Relatively few sugars are involved in the formation of the common polysaccharides. The most striking feature is that in many cases a given polysaccharide will draw on only one or occasionally two types of sugar molecule.

We see therefore that all these types of macromolecule are constructed by building up from sets of smaller molecules. For any given protein, nucleic acid or polysaccharide, how closely controlled is the order in which their constituent residues are arranged?

The answer is surprising and was indeed thought at one time to be unbelievable. It appears that in the proteins and nucleic acids there is almost complete control. Barring accidents, a protein containing many hundreds of amino acids or a nucleic acid containing many thousands of nucleotides will be turned out by the synthetic machinery of the cell time and time again, without alteration. One species of protein or nucleic acid is absolutely distinguished from any other by its amino-acid or nucleotide sequence. This tight control is essential because the functions of these macromolecules are sharply dependent on structure and even a small change in properties can be fatal to the delicate balance of physical and chemical events in living material.

The situation with the polysaccharides is quite different. The question of sequence does not arise in those polysaccharides which consist of a single sugar. Where more than one is used there is sometimes a measure of control of the sequence, so that if there are two, say, they may alternate along the chain. In other molecules where the desired function is less critically dependent on structure there may be less tight control.

Subsequent chapters will describe a little of what is known about the structure of these macromolecules, the way in which structure influences function, and the way in which the organism ensures that the correct structure is obtained during synthesis of the macromolecules. These chapters will seek to show that the remarkable properties which distinguish living matter from non-living are very largely a result of the properties of macromolecules. These derive in their turn from recognizable features of the structure. The development of life had to await the discovery of ways of ensuring that structures with beneficial properties could be repeated and safeguarded against deleterious changes from generation to generation.

Non-covalent interactions

The covalent bonds involved in the formation of the macromolecules have been mentioned and we must now consider the non-covalent ones. As will be seen in the appropriate chapters these are just as important to the structure and function of the finished product as the covalent bonds. In fact, a look at the role of these non-covalent forces gives some clue as to why biological macromolecules became necessary in the first place.

Little variation is possible in the strength, direction or other properties of any given covalent bond. On the other hand, the properties of non-covalent bonds depend far more on environment. The variety of biochemical reactions and the need for their control and integration would easily exhaust the versatility of available covalent chemical interactions, numerous though these are. The non-covalent interactions, which do have the required extra versatility, cannot be maintained in a useful form when small molecules are moving at random in free, dilute solution. However, when the interacting elements are anchored to one another in macromolecules the situation changes dramatically. Combinations of non-covalent forces can now be produced that are useful, powerful, and capable of precise and minutely graded variation.

These combinations of forces, and the local areas of high concentration of reactants that they make possible (pp. 47, 49), are the real foundation of the differences between biological and non-biological chemistry and thus of the differences between the living and the non-living state.

This point is rarely stressed in textbooks. Some feeling for its validity can be obtained from a glance at those macromolecular products of the living organism in which the rule is suspended and covalent bonds predominate. In such products there is a large number of covalent linkages between different parts of the molecule: they include horn and the exoskeleton of insects (p. 33), and hard (lignified) wood (p. 67). These are organic materials that we intuitively regard as the 'dead' parts. They change little on death of the organism or separation from it. Our intuition in this respect is a recognition of the suppression of the effects of weaker but more responsive non-covalent interactions by the stronger, inflexible, covalent bonds.

Potentially the strongest of the non-covalent interactions is the ionic bond, formed between ionized groups of opposite sign. In macromolecules most such groups are exposed to water (p. 12, but see p. 51 for an exception), and the ionic bond, like all electrostatic interactions, is screened as a result of the high dielectric constant of water and much weakened. The bond, therefore, is not now thought to be quite so important in biochemistry as it was.

The hydrogen bond is next in order of potential strength. A number of groups containing hydrogen, notably $-OH$ and $-NH_2$, have an unequal distribution of electrons in which the hydrogen atom has less of them than its share. Other groups which do not contain hydrogen, $>C=O$ for example, also have unequal distributions. The general term for a structure in which electric charge is unevenly distributed is *dipole*. A weak electrostatic interaction may be formed between two dipoles, of the type

$$\overset{\delta-}{-O}-\overset{\delta+}{H}------\overset{\delta-}{O}=\overset{\delta+}{C}\lessgtr$$

which, if at least one of the dipoles has a hydrogen atom, is called a hydrogen bond.

Although hydrogen bond-forming groups are common in macromolecules (which is doubtless no coincidence), the role of the bond, though of the greatest significance, has still been somewhat overestimated in the past, since it also is subject to interference by water. However, suitable groups do exist in regions of macromolecules from which water is excluded and here, though still weak, hydrogen bonds are numerous enough to make a vital contribution (e.g. p. 54) when taken together.

The hydrogen bond is just one of the types of force that arise through the interaction of dipoles. Many other types of charge inequality exist in molecules found in biochemistry and forces will arise between them.

Some dipoles, like those already mentioned, are a permanent property of the molecular structures involved, and the magnitude of the charge inequality (and thus the strength of the force) is constant with time. Other kinds of dipole arise from the random fluctuations in the density of electronic charge that occur around any atom, and here we can speak only of an average value for the strength of the forces that are produced. The charge inequalities of both permanent and fluctuating dipoles can themselves produce additional inequalities in nearby structures, and these resulting *induced dipoles* can then join in the formation of forces.

Individually, most of the forces that we described in the preceding paragraph are weak and are effective only at very short range. One might therefore think that they are unimportant. However, it would be wrong to dismiss them for two reasons. The first is that there are many dipoles in macromolecules and the resultant force obtained by summing all the dipole interactions can be considerable. The second reason is that mathematical analysis shows that in many cases such a resultant, even though it derives from short-range forces, is itself a *long*-range force. In fact the resultant of a large number of dipole forces falls off with distance more slowly than do forces obeying the inverse square law. Thus though perhaps weaker initially, the resultant may outreach more normal electrostatic forces.

The last type of force we shall consider is the hydrophobic interaction. As two oil drops coalesce when they touch in water, so adjacent hydrophobic structures (such as the non-polar side chains of certain amino acids for example) find that the closer they are together, the more stable the arrangement is (Fig. 2.4). The underlying theory of this common-sense conclusion is related to the hydrogen-bonding properties of water and will not be discussed here.

The hydrophobic interaction is of prime importance to the structure and function of macromolecules (as the next four chapters will show). It plays a major part in maintaining the structure of both

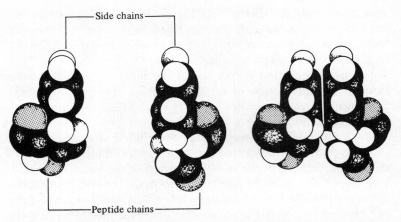

Fig. 2.4. The close approach of two hydrophobic side chains. Both residues are phenylalanine, and the benzene ring of each side chain is seen edge-on, pointing upwards. The peptide chains are represented by the cluster of atoms at the bottom of each structure. Carbon atoms (and the N in the peptide bond) are dark in shade, hydrogen atoms are light and oxygen atoms are grey.

proteins and nucleic acids. Both types of macromolecule have elements which favour hydrophobic bonding and elements which are hydrophilic. The hydrophobic elements will naturally lie, as far as is possible, in the centre of the structure, away from the solvent water. The hydrophilic elements will, on the contrary, lie on the surface of the molecule where they may interact with the water. We may draw an analogy with a detergent micelle (Fig. 2.5) with which you may be more familiar. The task of a detergent is to lift water-repelling dirt from soiled surfaces and permit it to be dispersed in water. Detergent molecules have both hydrophobic and hydrophilic regions. The hydrophobic regions associate with each other and with the dirt, and the hydrophilic regions interact with water in precisely the same way as do the corresponding elements of macromolecules. This explains how the dirt is dispersed, but more importantly from our point of view it accounts for the overall similarity of Fig. 2.5 to the generalized picture of a protein (Fig. 2.6) and a nucleic acid (Fig. 6.3).

As well as simply maintaining the structure, hydrophobic regions provide chemical environments quite dissimilar to those found in free aqueous solution. The unusual reactivity of certain key groups, in the catalytic site of enzymes, for example (p. 51), is a consequence of their being in such environments.

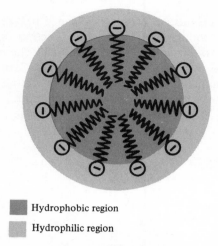

Hydrophobic region
Hydrophilic region

Fig. 2.5. A detergent micelle. The same dark and light tones will be used for other hydrophobic and hydrophilic structures throughout this book.

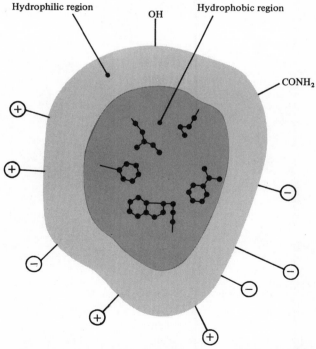

Fig. 2.6. The distribution of hydrophobic and hydrophilic regions in a protein molecule.

Thus, in biochemistry, water should not be seen simply as a passive solvent, though that is part of its function, nor yet just as a source of the protons and OH$^-$ needed for biochemical reactions, though that too is a part of its function (pp. 112 and 154), but as actively and uniquely concerned in maintaining the character and stability of the essential elements of living matter.

3 Protein structure

As we stated in the last chapter the proteins differ from one another in the selection and arrangement of their constituent amino acids. The next two chapters will show something of the many types of function undertaken by proteins. This remarkable range is achieved either solely by permutation of amino acids or with the addition of at most one or two other small molecules which are not amino acids (Table 3.2). We shall now examine the amino acids used in proteins in terms of the types of forces mentioned in the last chapter, and see what properties each brings with it to help in the task of producing a functioning molecule.

Proteins are of considerable size, usually having hundreds of amino acids, and they might therefore be expected to be sprawling, ill-defined structures. In fact the majority of proteins are compact, highly convoluted molecules with the position of each atom relative to the others determined with great precision. An error in position of a constituent part of as little as the diameter of one atom may be sufficient to inactivate a protein. Thus we have to consider more than just the few amino-acid residues that are directly involved in the interaction between the protein and other molecules. Many other residues will make as vital a contribution to function in an indirect way, by maintaining the required precision in the structure of the protein itself. Both types of contribution involve the covalent and non-covalent interactions mentioned in the last chapter, and a major part of this chapter will be devoted to examining the way in which these interactions are employed in the stabilization of the structure of the proteins. Their application to the external behaviour of proteins is best left to the chapters on protein function which follow.

Table 3.1 shows those amino acids that are normally found in proteins. (There are many other naturally occurring amino acids, but these do not appear in proteins.) All, with the exception of proline, which is further discussed below, have their amino and carboxyl groups joined to the same carbon atom (called the α carbon). All except glycine have at least one centre of asymmetry.

Table 3.1. *The side chains of the amino acids commonly found in proteins*

HYDROPHILIC

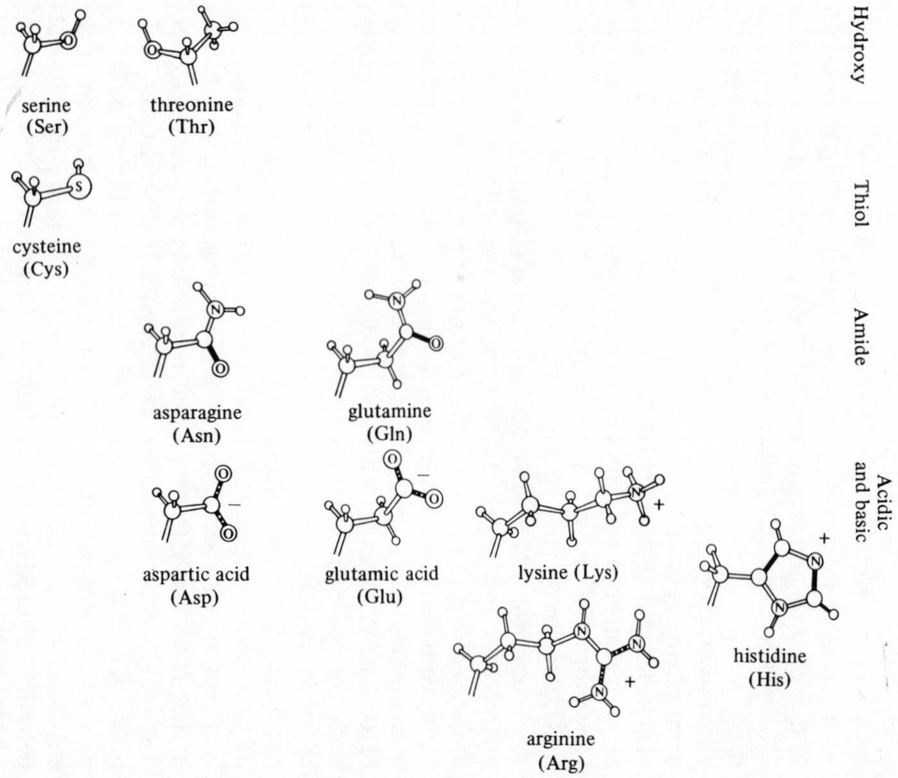

Hydroxy   Thiol   Amide   Acidic and basic

histidine (His)

lysine (Lys)

arginine (Arg)

glutamine (Gln)

glutamic acid (Glu)

threonine (Thr)

asparagine (Asn)

aspartic acid (Asp)

serine (Ser)

cysteine (Cys)

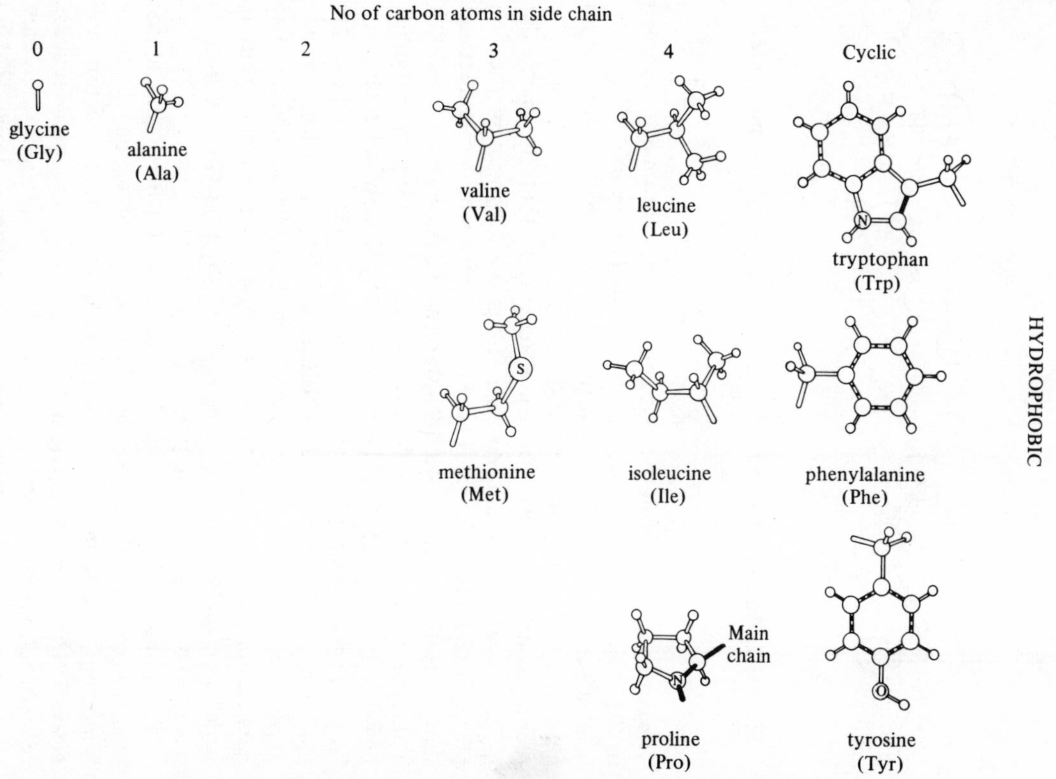

HYDROPHOBIC

No of carbon atoms in side chain

Cyclic

The three-letter symbols are those used in writing amino-acid sequences. Double bonds are shown as solid lines, and partial double-bond character is indicated by lines that are alternately open and solid.

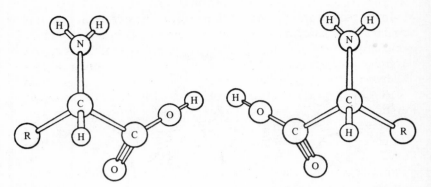

Fig. 3.1.1. An L-amino acid.   Fig. 3.1.2. A D-amino acid.

The configuration about the asymmetric carbon atom is that shown in Fig. 3.1.1 as opposed to Fig. 3.1.2. The two amino acids which have a second asymmetric centre in the side chain also show an absolute preference for one of the possible configurations at the second centre. We do not know why it is that the particular α-carbon and side-chain configurations were chosen in the first place, but it is characteristic of the chemistry of biological molecules that, once the choice was made, it was adhered to with great rigidity. This type of behaviour will be seen again with the sugars; it is made possible by the fact that the synthesis of biochemical molecules, large or small, takes place at *surfaces*. The surfaces referred to are mainly the surfaces of macromolecules, and since these contain asymmetric elements themselves they are able to discriminate between optical isomers. (By contrast, in synthesis in free solution, which is usually what happens in non-biological chemistry, such discrimination is largely impossible.)

The amino acids in the Table are classified according to the characteristic of the side chain that appears to be the most important in determining its contribution to protein structure and activity. Members of the first group show a graduated increase of hydrophobic property as the size of the side chain increases. (Proline is in a class of its own since it is not truly an amino acid but an *imino* acid. The peptide bond formed by the $>$NH group of proline will clearly have geometrical and chemical properties which differ from those of the peptide bond formed by an –NH₂ group. A consequence of this is, for example, that the α-helix (p. 21) cannot accommodate proline residues.)

In the second group, those having hydrophilic side chains, there is again a graduation of character. Aspartic acid is more strongly acidic than glutamic acid, histidine is a weaker base than lysine but arginine is a stronger one. Also, there is a spread of more generally hydrophilic, hydrogen-bonding properties. Cysteine (pronounced SISTAY-EEN) is in a class by itself in that it may be converted to cystine (SIS-TEEN) by oxidation (see Fig. 3.2, below).

It must not be thought that the characteristics mentioned for each group are exclusive to that group or are its sole characteristics. Several amino acids not in the hydrocarbon group have some hydrophobic properties in parts of their side chain, e.g. the aliphatic hydrocarbon part of the lysine side chain (see p. 48 for an example of how this dual nature can be exploited). Another example is the ability of the –OH group of tyrosine and the –SH of cysteine to ionize. Several side chains are used in proteins for their chemical reactivity – cysteine and serine in particular.

As we saw in the last chapter, the incorporation of amino acids into polypeptides takes place by condensation and the formation of peptide bonds. The sequence of incorporation of amino acids, the so-called primary structure, is of absolute importance in determining the polypeptide's properties. (Experimental methods exist for determining the sequences of proteins, and we give one such sequence on p. 61.) There are about 10^{195} possible sequences for a protein of 150 residues, but every molecule of a given species of protein always has exactly the same sequence, and hence exactly the same properties.

We have already remarked that the properties of a protein depend on particular juxtapositions of side chains, with the consequent interplay of the different types of characteristic mentioned. This juxtaposition is not merely one-dimensional, along the chain. The chain exists in three dimensions and it is its convolution to a precisely determined shape (tertiary structure) that allows the full interplay of characteristics of side chains. Proteins, which are therefore quite compact, are often divided into the categories 'globular' and 'fibrous'. An absolutely spherical shape is not required for a protein to be classed as globular, but when the ratio of length to width (known as the axial ratio) reaches or exceeds approximately 5 : 1 it is classified as fibrous.

Interactions stabilizing protein structure

The most important covalent interaction results from the oxidation of the cysteine residues to form cystine. This allows the joining or looping of polypeptide chains without the use of the peptide bond (see Fig. 3.2). These so-called disulphide bridges are frequently found in proteins as a general aid to the stabilization of structure, and they are also used where special mechanical properties are required (e.g. see p. 33). In exceptional circumstances cross-linking is brought about by other types of covalent bond, and these too are mentioned on p. 33.

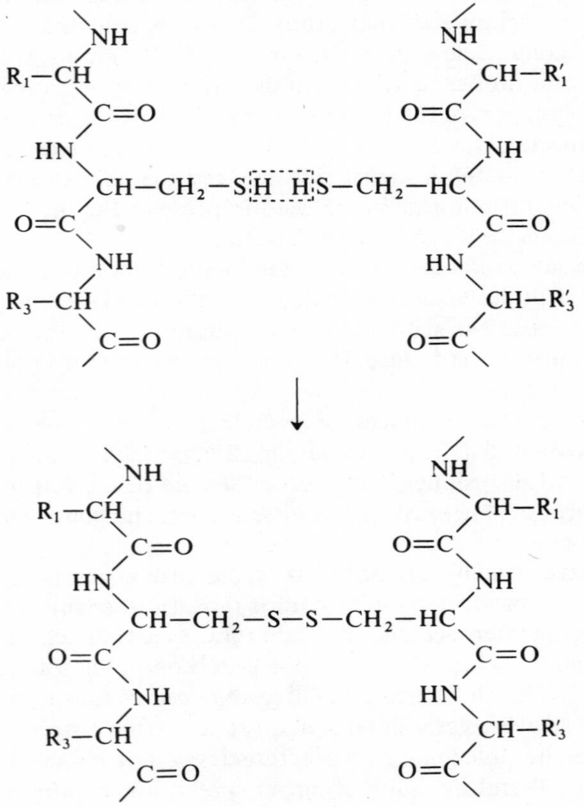

Fig. 3.2. Formation of the disulphide bridge.

The ionic bond can occur by the interaction between on the one hand the positive charges on histidine, lysine, arginine and the α-amino group, and on the other the negative charges of aspartic acid, glutamic acid and the α-carboxyl group. They are less frequently used in the stabilization of protein structure than, say, hydrogen bonds (but see Fig. 3.6). As will be seen later, they are more frequently found in interactions between proteins and other molecules.

The hydrogen bond is found between the side chains of the members of the third group of amino acids in Table 3.1. Moreover we may note that the peptide backbone itself is capable of a more active role in stabilizing the structure than simply holding the amino acids together. It has $>$CO and $>$NH groups regularly disposed

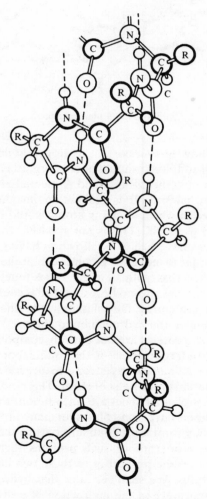

Fig. 3.3. The α-helix.

along its length and these groups are eminently capable of hydrogen bonding to each other. As the groups are regularly arranged, it is not surprising that hydrogen bonding between them can give rise to regular structures. Features of this type are described as *secondary structure* and two, the α-helix and β-pleated sheet (Figs. 3.3 and 3.4), are found in most proteins.

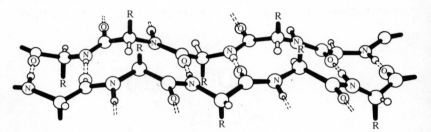

Fig. 3.4. The β-pleated sheet.

The Figures show how important hydrogen bonding is in these structures, but in addition it should not be forgotten that, viewed as collections of atoms, the α-helix and β-pleated sheet are extraordinarily tightly packed. It is difficult to imagine many other regular structures in which so many atoms would be placed close enough together to touch. (This is not apparent from the Figures because the atoms have had to be shown as having less than their proper radii in order to make the pictures comprehensible at all.) It is often overlooked that, with so many atoms touching, the short-range dipole forces can also make a substantial contribution to the stability of these structures; they should not be thought of, therefore, as depending on the hydrogen bond alone.

The reason why proteins are not entirely composed of elements with regular secondary structure is that interactions between side chains will often be sufficiently strong to override the tendency toward the formation of the helix or sheet. This is not to say that the α-helix and the β-pleated sheet are uncommon: they are used as reinforcing members (struts and plates) in many proteins (Fig. 3.5). In a few cases they form a predominant part of the structure of a protein and the properties of such proteins then depend to a significant extent on the properties of these two kinds of ordered secondary structures. We shall see later that hydrogen bonds are also of great importance in the interaction of proteins with other molecules.

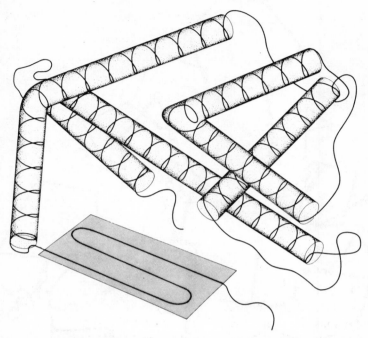

Fig. 3.5. Helical and sheet-like regions in a globular protein.

Other dipole forces must certainly be present throughout protein molecules since so many elements of the structure produce dipoles. They are thought to be quite important, especially in stabilizing quaternary structure (p. 25). Because of the large numbers of relatively feeble interactions which are involved little is known in detail of their contribution.

The hydrophobic interactions are probably the most powerful in stabilizing the structure of proteins, and are also prominent among the forces involved in their interactions with other molecules (see p. 37). Now that X-ray crystallography has made it possible to look at the tertiary structure of proteins it can be seen just how important they are (Fig. 3.6). As suggested previously the molecule resembles a detergent micelle. The majority of the hydrophobic elements cluster together at the centre, with only a few of them exposed to the aqueous solution. The hydrophilic elements, on the other hand, are almost all exposed.

Some proteins consist of aggregates of (often) similar protein subunits. This so-called quaternary structure should not be thought

Fig. 3.6. A computer-drawn diagram (with perspective) showing some of the types of interaction stabilizing the conformation of the enzyme lysozyme (see also Fig. 5.5.1). Selected parts are shown of the polypeptide backbone of the protein and a very few of the side chains. Note the disulphide bridge (the cystine residue: one of four in the protein); the ionic bond between the NH_3^+ of the lysine and the COO^- of the carboxyl terminus of the protein (the only ionic bond in the protein); the close approaches between hydrophobic residues (just a few of those occurring in the complete structure). The clearest of the hydrophobic interactions is probably the 'sandwich' of a methionine side chain between two tryptophans. For clarity, the atoms are not drawn to their full size; if they were it would be seen that many of the side chains are virtually touching.

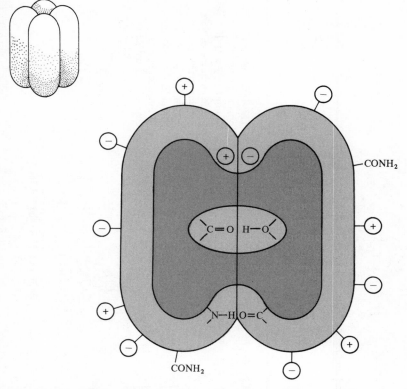

Fig. 3.7. Idealized section through a tetrameric arrangement of protein molecules showing types of interaction at an interface between two of the monomers. The arrangement is indicated in the inset drawing. Note that structures of this kind can have a central channel. The plane of section for the main drawing does not pass through the centre of the tetramer, but through the centres of two of the monomers.

of in terms of a random aggregate. The number of subunits involved and their geometry is precisely controlled. Some proteins rely on the extra possibilities conferred by a quaternary structure for the full expression of their activity (see pp. 42, 62 and 238). Quaternary structure is stabilized once again by the full range of non-covalent forces (see Fig. 3.7).

Denaturation

Proteins, as they are found in living tissue, are usually freely soluble materials with some definite chemical or mechanical function.

Table 3.2. *Co-factors and prosthetic groups*

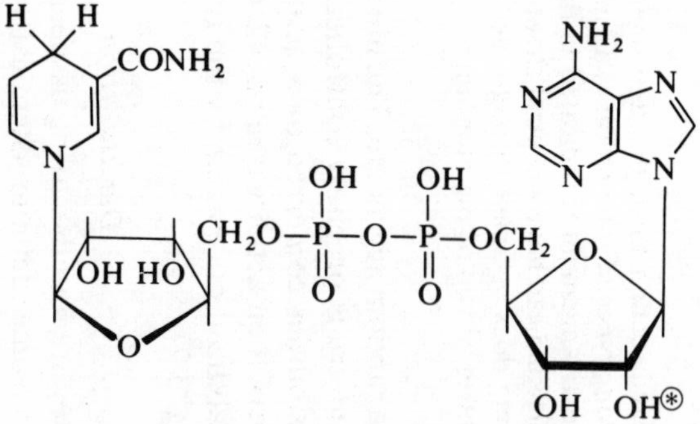

nicotinamide adenine
dinucleotide-reduced
($NADH_2$)
(see pp. 104–5)

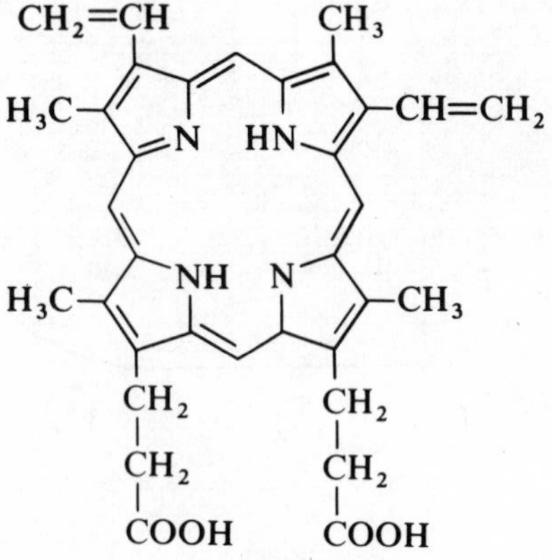

NADP, NADPH₂ have Ⓟ at ✳

porphyrin derivatives
(see pp. 39–43 and 109–10)

This structure is the
porphyrin which
(complexed with Fe^{2+}) is
found in haemoglobin.
Similar structures (in
which $Fe^{2+} \rightleftharpoons Fe^{3+}$) are
found in the cytochromes.
Other porphyrins
(complexed with Mg^{2+})
occur in the chlorophylls.

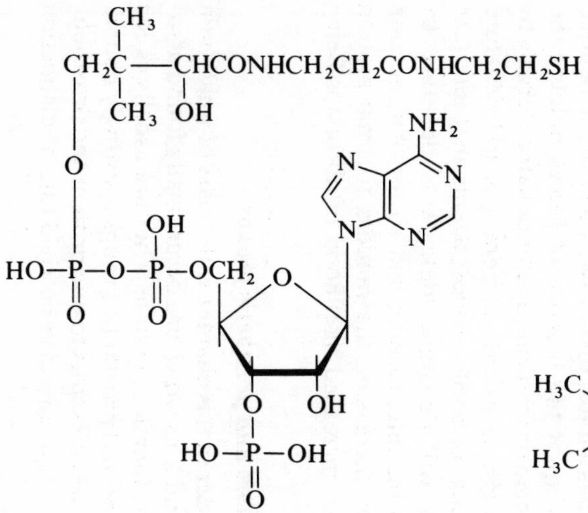

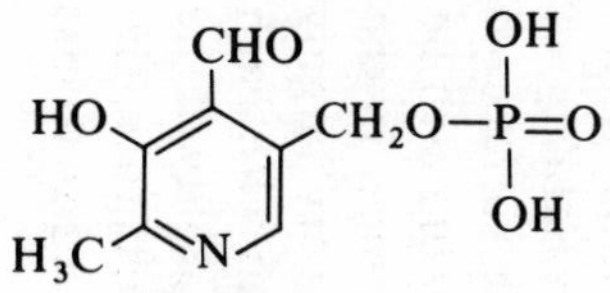

pyridoxal phosphate
(see p. 172)

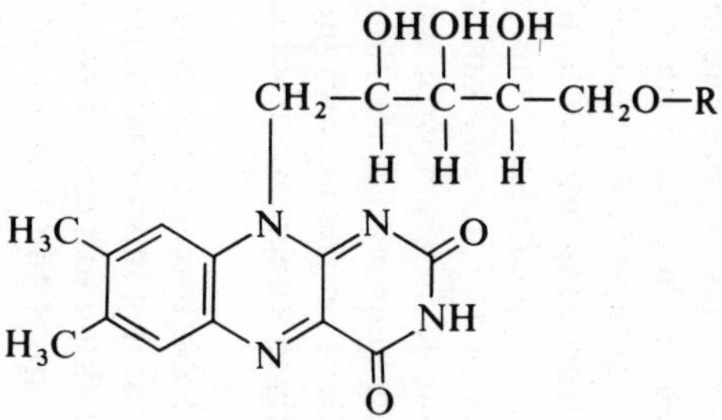

R = H: Riboflavin

R = (P): Flavin mononucleotide (FMN)

R = -(P)-(P)-ribose-adenine: Flavin adenine dinucleotide (FAD)

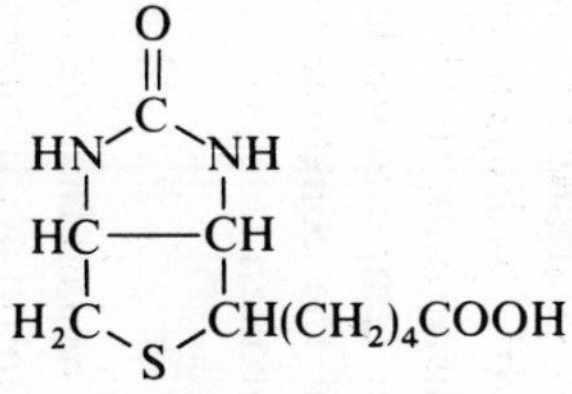

riboflavin and derivatives
(see pp. 111)

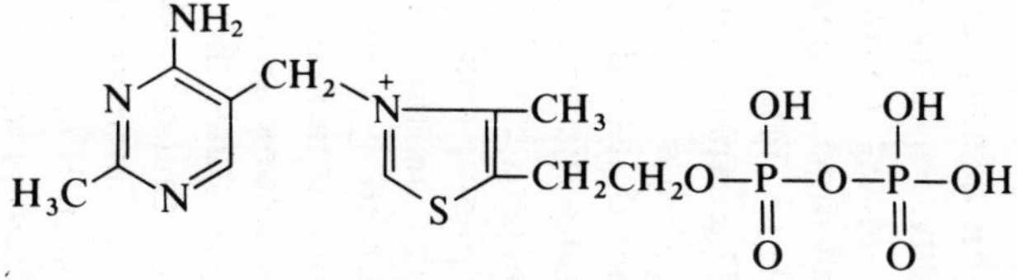

thiamine pyrophosphate (TPP)
(see pp. 133–4)

coenzyme A (CoASH)
(see pp. 133–4)

biotin
(see p. 167)

When harshly treated they become far less soluble and lose the ability to carry out their function. They are then said to be *denatured*. Curdled milk and boiled egg-white are examples of denatured protein as, in a sense, is the major constituent of blood clots (p. 34).

The explanation for the changes in properties when a protein is denatured is that the tertiary structure has been disrupted. Thus any agent that leads to the weakening of any of the interactions maintaining the tertiary structure will be a denaturant. If the tertiary structure is unfolded, residues which are intended to join in hydrophobic and other interactions *within* the molecule find themselves on the surface. Here they have an equal probability of interacting with similar residues now exposed on the surface of adjacent molecules. The resulting intermolecular network will be less soluble than the assembly of native molecules, since, in the native state, the residues of the micelle surface are very largely those that are more prone to interact with water than with each other. Since, as has already been stated, the functional ability of the protein is closely dependent on the correct tertiary structure, the unfolding will lead not only to a decrease in solubility but also to loss of biological activity.

With our knowledge of the forces maintaining tertiary structure we can now explain the action of individual denaturants. Proteolytic agents can denature by making just one or two nicks in the peptide backbone if these nicks upset the balance of forces holding the molecule together; reducing or oxidizing agents denature largely by the cleavage of the disulphide bridge; extremes of pH denature by charging or discharging ionizable groups; agents that interfere with hydrogen bonding will denature; detergents denature by disrupting the hydrophobic interactions; and so on. An increase in temperature leads, by increasing the random thermal motion of the constituents, to an increased likelihood of all bonds being broken.

Co-factors and prosthetic groups

The last feature to consider in this chapter is the class of small (non amino acid) molecules and ions which are occasionally found associated with a protein in its functioning state. They are usually called co-factors if weakly bound and prosthetic groups if strongly bound. (Certain substances may be co-factors to one protein and prosthetic groups to another. There is in any case no rigid line of distinction between the two terms.)

There are tasks which are beyond even the large range of abilities of amino-acid side chains, and it is essential for other molecules to be used on occasion. Co-factors and prosthetic groups, bound in sites specifically intended to enhance their usefulness, may behave in a way which would not be expected from their properties in free solution. This is analogous to the situation existing for the amino-acid side chains themselves. The protein therefore accommodates the co-factors in two ways. It provides a specific binding site and it controls the chemical environment of that site.

Table 3.2 lists the more common co-factors and prosthetic groups. It will be seen that some of the structures given are quite complex. Many organisms, man included, are not able to synthesize all of these structures for themselves, and they must then be obtained in the diet. Many vitamins and so-called trace elements fall into this category. The reason why only very small quantities of these substances are required is that the proteins with the require-ment for the co-factor or prosthetic group are usually present themselves in only catalytic amounts. We can contrast this situation with the dietary requirement for the amino acids themselves. In man, for example, eight of the protein amino acids cannot be synthesized and these have to be obtained in the diet. There is no question here of a need for trace amounts at a specific site: instead there is a general requirement for substances that are constituents of all proteins, so that considerably larger amounts must be ingested.

4 Protein function I

We will now see how it is that the same forces, acting between the same set of elements, manage to produce such a wide range of behaviour.

We may conveniently divide the proteins into three classes, depending on the way in which they contribute to the activities of living matter. The first class consists of the food storage proteins, the second of proteins which have a structural or mechanical role and the third of those that act by binding to other molecules.

Food storage proteins

These may be quickly dealt with. In order to synthesize a protein an organism must have an adequate supply of amino acids. In some cases it may be able to meet all its requirements by its own efforts, making use of the biosynthetic pathways described in Chapter 17. Should demand exceed this internal supply, or should an organism be incapable of synthesizing a particular amino acid (e.g. see p. 184), an external source is necessary. In such cases food proteins may be consumed, and then digested to amino acids by means of hydrolytic enzymes. Almost any protein may serve as a source of amino acids in this way, but specialized molecules have been developed for particular purposes. For example, mammalian young are dependent on their mothers' milk for, among other things, a supply of amino acids to meet the severe nutritional problems posed by their early growth. This demand is met by a number of milk proteins, in particular by the caseins. Caseins, which are soluble until ingested, are found to be exceptionally easily denatured in the gut. This denaturation exposes many peptide bonds which are susceptible to enzyme-catalysed hydrolysis and the protein is readily digested.

We shall meet an instance below (p. 42) that shows how difficult it is to know when one has come to a full understanding of the function of a protein. However, as far as is known, this ease of denaturation and digestion is the major functional requirement for the casein molecule.

Structural and mechanical proteins

Elastic and inelastic structures

We consider now those proteins that act as more or less inert structural elements. These proteins are frequently of the high axial ratio (fibrous) type. They may be required to be elastic (for example, skin or hair) or relatively inelastic (for example, silk). As you might expect in the light of what we have said earlier on, elastic structures tend to be those in which non-covalent bonds predominate among those that resist stretching, while in rigid structures it is the covalent bonds that predominate in this role.

α-Keratin is a good example of an elastic structure. It relies for both its strength and its elasticity on the α-helix, one of the regular arrangements of peptide backbone hydrogen bonds (Fig. 3.3) mentioned in the last chapter. It will be seen that when the α-helix is stretched the bonds, although weak individually, maintain a spring-like structure in which they are parallel and thus may be expected to have some collective strength (Fig. 4.1.1).

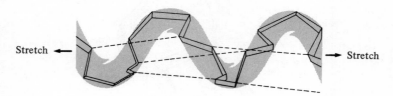

Stretch ←   → Stretch

Fig. 4.1.1. Stretching an α-helix.

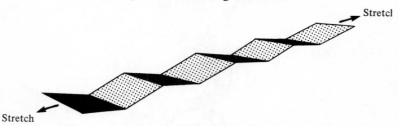

→ Stretch

Stretch ←

Fig. 4.1.2. Attempting to stretch a β-pleated sheet.

The inelastic protein α-fibroin occurs in certain silks. It typifies the β-pleated sheet structure (Fig. 3.4), another of the backbone hydrogen-bonded structures. Here the hydrogen bonds are at right angles to the direction of stretching and simply hold the bundle of adjacent polypeptide chains together. Extension is resisted by the full strength of the covalent bonds which in this case lie not perpendicular to but along the direction of stretching (Fig. 4.1.2).

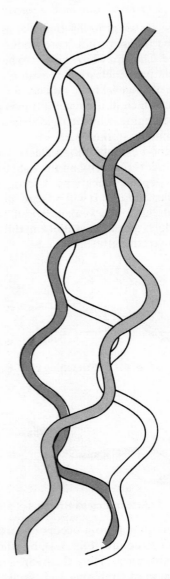

Fig. 4.2. The triple helical structure of collagen.

Collagen is another relatively inelastic fibrous protein. It has a remarkably high tensile strength and, weight for weight, is about as strong as steel. It has a complex structure based on a helical pattern (Fig. 4.2).

Elasticity or rigidity may be enhanced by covalent cross-linking of the protein chain. A relatively open network of cross-links confers elasticity while a tighter, more numerous, system of cross-links confers rigidity (Fig. 4.3). The covalent bonds used may be S–S bridges (p. 20) or may arise in other ways, including the exploitation of special, aromatic free-radical reactions initiated by enzymes, which form carbon–carbon bonds between side chains.

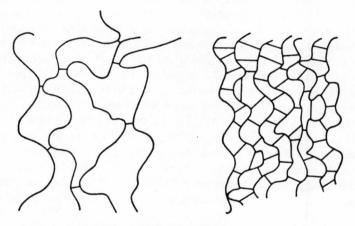

Fig. 4.3. Cross-linking of peptide chains.

In hair, the natural elasticity of α-keratin is enhanced by embedding it in a matrix consisting of proteins that are cross-linked by a moderate number of S–S bridges. In horn, another keratin-containing structure, the matrix has more S–S bridges and is rigid. The final event in the hardening of the blood clot (p. 34) is the formation of S–S bridges and peptide-like links between side-chain amino and carboxyl groups. Certain elastic tendons, which rival in their properties the most advanced synthetic rubbers, do not have a matrix but use the aromatic free-radical method for direct covalent cross-linking of the chain.

These free-radical reactions are also used in the formation of rigid structures. Insect cuticle, for example, consists of protein that is extensively cross-linked in this way. (It was only by developing this method of cross-linking that insects were able to develop a light exoskeleton which was nonetheless sufficiently rigid to sustain the stresses of flight.) Rigid structures may also be obtained by embedding the protein in a matrix of some material other than protein. The best known example is bone, in which the matrix is mineral. Crustacea, which have never achieved flight, accomplish the hardening of their exoskeletons by mineralization. The weight of mineral required to strengthen the exoskeleton is probably sufficient in itself to prohibit flight.

Muscle proteins

The principal structural components of muscle cells are bundles of fibrous proteins which include two called actin and myosin. Muscular contraction occurs when the bundles slide between one another, diminishing the size of each muscle cell and thus of the muscle as a whole. This sliding is thought to be the result of conformational changes of the proteins making up the bundles. The changes are almost certainly mediated by the transfer to the protein surface of the chemical energy inherent in a certain bond in the energy-storage compound ATP (see p. 101). The balance of forces is altered and this alters the conformation of the protein. This occurs a sufficient number of times for the microscopic changes in conformation to become visible as the gross movement of the limb of an animal.

Although myosin is in most of its characteristics a typical fibrous structural protein, it is unusual in that it also possesses subunits of its quaternary structure that have enzymic activity. The enzymic activity in question is the catalysis of the breakdown of ATP and is clearly connected with the contraction mechanism.

Fibrin and blood clotting

Fibrin is an interesting structural protein. In its native form (fibrinogen) it is a soluble protein of considerable axial ratio (see p. 19). When it is necessary to form a blood clot, two small breaks are made in the peptide chain by an enzyme; the balance of forces stabilizing the fibrous structure is thus disrupted. A shift occurs to a more globular structure (fibrin) and residues previously buried come to the surface. Many are able to join in hydrophobic and other interactions between adjacent protein molecules and stabilize a great number of intermolecular links – the analogies with the denaturation process are obvious. Thus the solubility is lowered and a clot forms. As already mentioned, the final event is the cross-linking of the molecules by covalent bonds.

This phenomenon of activation by removal of a part of the chain is met with elsewhere. It occurs when it would be an embarrassment to have a protein expressing its full function before it was needed. Examples are some hormones, and enzymes which degrade cellular constituents.

Structural proteins of low axial ratio

Globular proteins are sometimes used to solve structural problems, although not so frequently as are the fibrous proteins. An example of the use of globular proteins is the encapsulation of virus nucleic acid into a rigid protective quaternary structure (see p. 62).

Proteins that bind other molecules

We come now to the third and most intensively studied class. These proteins exploit to the full the possibilities of combination of the different amino-acid side chains. They do so to produce a site on their surface which has both a specific shape and a specific array of forces. This site will bind a particular molecule or part of a molecule with great tenacity. This is because it is a perfect fit, both in the geometrical sense, and in terms of the chosen forces meeting just those parts of the structure to be bound upon which they can best act. Thus the binding is both powerful and reasonably specific, since even a small change in the structure to be bound would be likely to spoil the fit and upset the interplay of forces (see also enzyme specificity, p. 44).

If the sole function of the protein is to bind, as is to some extent the case with haemoglobin and with the immunoproteins (see below), the matter ends there. In some cases, however, it is necessary not only to bind a molecule but to provide a special environment in which to modify its properties. The most important examples of this are the enzymes (Chapter 5), which enhance the reactivity of the molecules which they bind.

Table 4.1 shows a convenient classification of the better known types of binding protein.

We propose to discuss just one of these proteins, haemoglobin, in detail, as an example of all the others. However a few notes on Table 4.1 may be helpful.

The hormones, whether or not they are proteins, are thought to influence metabolic rates by their ability to bind at a specific site. This site may be on a *membrane*, which has its properties modified as the result of the binding. The rate of passage of metabolites through the membrane might then be significantly altered and this

Table 4.1. *Proteins that bind to other molecules*

Hormones*	Transmit instructions for the control (p. 238) of the levels of metabolic activity
Immunoglobulins	Bind to and inactivate foreign materials invading the body
Enzymes	Catalyse biochemical reactions
Carrier proteins	Transport molecules, ions or electrons from one place to another either within the cell or over greater distances within the organism

* Not all hormones are proteins.

would obviously influence the rate of metabolic activity within the cell or other structure bounded by the membrane. Alternatively, a hormone may bind to a specific *enzyme* and induce a change in the tertiary structure which alters the catalytic activity. We shall see (Chapter 23) that a change of activity of just one enzyme can frequently influence the rates of a large number of biochemical reactions. Finally hormones may bind to the *genetic apparatus* and influence the rate of synthesis (p. 215) of an enzyme. To change the amount of an enzyme in a cell is equivalent in many ways to changing its activity and will have similar consequences.

Insulin is a well-known protein hormone produced by certain specialized cells in the pancreas. It is an example of a protein that is produced as an inactive precursor (pro-insulin) which is shortened to the active form by specific proteolytic cleavage on secretion. In its active, circulating form it is practically the smallest protein known, having only fifty-one amino acids. It has profound effects on, in particular, carbohydrate metabolism. It is discussed further on p. 243.

The immunoglobulins deal with invading substances in the body and play a major part in such phenomena as immunity and transplant rejection. The system works in the following fashion. Mechanisms exist by which the body is able to recognize the presence of certain classes of alien molecules known as antigens. (Antigens are usually macromolecules: protein, nucleic acid, polysaccharide, lipid or a combination of any of these.) Once the presence of the antigen is recognized, the synthesis of the immunoglobulin (antibody) begins. A large number of different antibodies are produced in response to any one antigen. All contain a binding site (see above), which will fit and bind to one of the structural

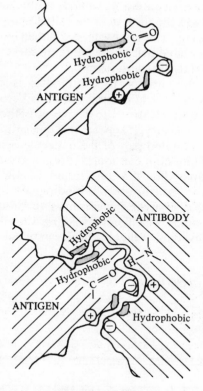

Fig. 4.4. Antigen–antibody combination.

features of the antigen that is not found in any of the molecules belonging to the host organism. For instance, suppose that there is an area of the invading antigen with the shape and combination of forces shown in Fig. 4.4.1. Somehow an antibody is selected from the many designs avilable to the organism with a binding site nearest to that which would provide a perfect fit (Fig. 4.4.2) in both of the senses used on p. 35. Since it appears that an organism has millions of possible structures at its disposal, the best of these is, for most conceivable antigenic features, likely to be a very good fit indeed. If this is so the antibody will bind so strongly to the antigen as to make the association effectively irreversible.

The antigen thus complexed is likely to lose its capacity to act. A bacterium or tissue transplant covered in antibodies will probably cease to function. Immunization relies on the fact that having once learnt to synthesize an antibody the body is able to produce it very much more quickly should it ever be called upon to do so again. Rapidly multiplying viruses or bacteria will then have less chance of gaining the upper hand before the antibody concentration reaches an adequate level to deal with them.

The protein structure of the antibodies is currently under investigation. A great deal of information is becoming available, and the study of antibody structure is now one of the most promising areas of protein chemistry. It appears that one antibody differs from another only in a limited part of the molecule (which presumably includes the binding site); considerable regions of the structure are much less variable (Fig. 4.5). Attention is now turning to the elucidation of the structure of the binding site itself, and to the means by which the organism contrives to produce only those particular antibodies which have binding sites that will fit the antigen that is invading at that particular moment.

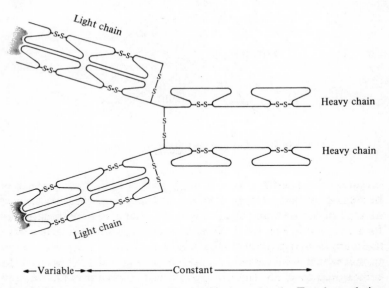

Fig. 4.5. The overall structure of an antibody molecule. Two long chains (called 'heavy') and two short chains (called 'light') are linked together by disulphide bridges. The parts of the molecule that tend to be constant in amino-acid sequence from one antibody to the other are distinguished from those parts that are very variable, and which are responsible for the creation of the two antigen-binding sites. (The binding of the antigen at each of these sites is suggested by the shaded areas at the left of the figure.) The antibody molecules also possess some fairly short carbohydrate chains (not shown).

The enzymes, catalysts of biochemical reactions, are among the most important constituents of living matter. They are considered in a separate chapter (Chapter 5).

Some carrier proteins exist which transfer molecules, ions or electrons over quite small distances only. Thus the membranes of the cell have some of the properties of the semipermeable membranes found in ordinary chemistry but add to them a number of remarkable features for which it is believed that protein–lipid complexes are responsible. In particular, biological membranes, unlike ordinary semipermeable membranes, can bring about the transport of solute molecules into areas of *higher* concentration. Biological membranes show great specificity in terms of which molecules they will transport and in which direction they will transport them. Membranes exist, for example, which will accept D sugars and reject L sugars, while certain nerve membranes will concentrate Na^+ ions on one side and K^+ ions on the other. We discuss these matters in a little more detail in Chapter 21.

The cytochromes (Chapter 8) transport electrons over short distances in the cell, and many enzymes may be said to have short-range transport functions as well as catalytic ones. This is because in several metabolic pathways (see Chapter 9) the enzymes concerned are arranged in close and defined proximity to each other in assemblies built into the membrane. The speed at which some of these assemblies operate is such that there would not be enough time for the reactants to diffuse in free solution from one site to another; they must be passed directly from one enzyme surface to another.

Other protein carriers exist to transport a necessary substance from one part of an organism to another quite distant part. These may frequently act as stores, holding the substance carried until it is required. The leading example of such proteins is probably haemoglobin, an oxygen carrier (see below). There are also several proteins which carry particular metal ions about the body as well as many which carry specific small molecules. All rely on the usual covalent and non-covalent forces to provide an area of high affinity for the substance to be carried.

Haemoglobin: biological phenomena explained at the molecular level

In order to show in the greatest possible detail how the principles of Chapter 3 may be applied to explain the activity of proteins we have chosen two examples. One is the enzyme lysozyme, which will be

dealt with in the next chapter, the other is haemoglobin. Haemoglobin is chosen to represent the non-enzymic proteins partly
because of its medical and biological significance and partly because
it presents one of the most rewarding instances of the study of
biology in molecular terms.

The site of interaction between oxygen and haemoglobin is an
atom of iron (Fe^{2+}). It is prepared for its role in oxygen binding by
being held in an environment containing five nitrogen atoms at
precisely the correct orientations and distances to allow co-ordination to occur. The positioning of the nitrogen atoms is controlled by
having four of them bound together in a rigid ring system of the
correct dimensions (the porphyrin ring, see Fig. 4.6). This molecule

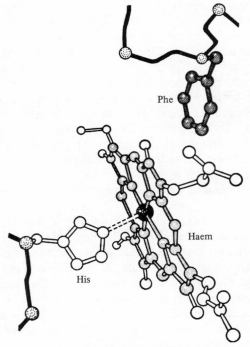

Fig. 4.6. Haem–protein interactions.

has several hydrophobic areas, and these and other interactions are
used to fix the ring on to a suitably shaped, partly hydrophobic patch
on the protein molecule. The nitrogen of a histidine side chain takes
up the fifth co-ordinating position. The sixth position, which is

called for by the geometry of co-ordinated iron, is found to be particularly suitable for binding oxygen. This binding is strong enough for the carrier function to be fulfilled but not too strong to prevent the oxygen from being given up when required. (Carbon monoxide is poisonous because it will bind in the same place but much more strongly and will thus not make room for oxygen when required. This phenomenon is analogous to competitive inhibition in enzymes (p. 48).)

Haemoglobin, which is found in many groups throughout the animal kingdom, has been developed by the vertebrates in particular into an extremely subtle instrument. It is now possible for us to appreciate how this has been done, since the structure of the molecule has been determined in atomic detail. It is very much in the interests of an animal to have an oxygen carrier which does not have a straight-line relation between the amount of oxygen available (or required) and the percentage oxygenation of the carrier, but a so-called sigmoid relationship (Fig. 4.7). The advantage of the sigmoid curve is that almost the full oxygen capacity of the carrier can be called on with a smaller relative fall in the oxygen concentration. If a straight-line relationship existed then the carrier would still be hanging on to some of its oxygen at very low oxygen levels, and would still be inclined to give up some of its oxygen at very high oxygen levels. This greater efficiency allows the vertebrate cell to exist in an even environment without the great fluctuations in oxygen concentration which would be necessary to utilize the capacity of a less sophisticated carrier to the full. These advantages

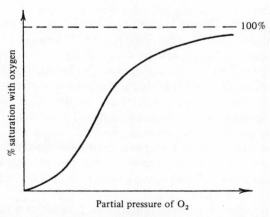

Fig. 4.7. The oxygenation–deoxygenation curve of human haemoglobin.

must be a contributing factor to the evident success of the vertebrates as a group.

In the vertebrates as in the invertebrates, the active part of
haemoglobin is still the porphyrin ring (Fig. 4.6). The improvement
in performance is managed by means of a new refinement in the way
that the protein's side chains influence its environment. What has
happened is that four protein molecules have aggregated into a
specific quaternary structure (Fig. 4.8). This structure is so arranged

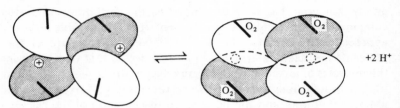

Fig. 4.8. Positively charged side chains revealed by the change in shape
following deoxygenation of haemoglobin (idealized).

that when one ring is oxygenated, the disturbance to the protein
structure at that point is transmitted by rearrangement of the
amino-acid side chains to the site of another porphyrin ring. The
consequent change in the environment of this second ring is such as
to increase its affinity for oxygen. Hence, once one oxygen molecule
has bound to the tetramer, another is much more likely to follow.
This gives the up-turn in the oxygenation curve which is required to
produce the sigmoid shape. The phenomenon of activation (or
deactivation) of a site on a quaternary structure by binding of a
molecule at a distant site is also exploited in the regulation of the
catalytic activities of enzymes (see p. 238). Haemoglobin has another important biological activity which could easily be overlooked;
it is a major source of the buffering power of blood. The dissociation
equilibria of its ionizable side chains (many of which project into the
solution) control pH very effectively in the same way as the
buffering substances used in non-biological chemistry. Even more
subtly, the molecule is arranged so that, when the tetrameric
structure is disturbed by oxygenation or deoxygenation, different
residues are exposed to compensate for the pH changes which
would result from the production of carbonic acid when the oxygen
is used up (Fig. 4.8).

The structure has some remarkable properties and it is clear
that even small variations would be likely to have unfavourable

results. A number of mutations (p. 184) are known which have given rise in certain individuals to haemoglobins in which one amino acid has been substituted for another. If the change, as is often the case, results in impairment of function, the individual will suffer from an anaemia. The precise clinical symptoms of this anaemia will depend on the nature and severity of the impairment of function of the molecule. Many anaemias which were first known solely as clinical patterns, sometimes of great complexity, have now been explained completely in molecular terms – a hopeful augury for other diseases. By knowing the location of particular amino-acid substitutions in the haemoglobin of the patients concerned, it is frequently possible to predict the alteration in the properties of the protein from the nature of the substitution. The consequences of this alteration may then be related to the nature of the disease.

In concluding this chapter we must stress that haemoglobin has been selected solely as an *example* of the way in which it is becoming possible to explain quite complex biological phenomena in terms of a few relatively simple ideas. Information of this sort, though possibly a little less complete, exists for a number of other proteins, and much more is likely to become available in the near future.

5 Protein function
II – the enzymes

The enzymes are the catalysts of biochemical reactions. No substance with the properties of an enzyme has been found which is not a protein – although, of course, many enzymes employ co-factors or prosthetic groups. It is now agreed that should such a substance ever be found, it will not be called an enzyme.

The enzymes are of particular importance as catalysts of most of the reactions that occur in cells. Their catalytic power is prodigious – frequently several orders of magnitude better than the corresponding non-biological catalysts. For example, one molecule of catalase, an enzyme which destroys hydrogen peroxide, is able to deal with over 1 000 000 molecules of H_2O_2 per minute. This represents an acceleration of at least 10^{14} times over the rate of the reaction without a catalyst. It is also many times better than the non-biological catalysts of the reaction, such as finely divided platinum. In fact, catalase depends for its action on an atom of iron, which is activated by the environment specially produced for it by the amino-acid side chains nearby. The extent of this activation is shown by the fact that 1 mg of iron in catalase is as effective a catalyst of the decomposition of H_2O_2 as is 10 metric tons of inorganic iron. It is naturally of the greatest interest to try to discover how amino-acid side chains can accomplish such effects, and, as we shall see later in this chapter, we have by now obtained a great deal of knowledge about such questions.

The specificity of enzymes is equally striking. We have already mentioned a case in which living processes distinguish between different optical isomers, almost the ultimate problem of ordinary preparative chemistry. This is but one example of the ability of enzymes to discriminate between possible substrates, however closely related they may be. Urease is an example of very high specificity. It vigorously catalyses the breakdown of urea according to the equation

$$CO(NH_2)_2 + H_2O \rightleftharpoons CO_2 + 2NH_3$$

Many compounds exist in which substituents are placed on the $-NH_2$ groups of urea. None are touched by the enzyme.

Another group of enzymes has a somewhat broader specificity, for example the digestive enzymes that catalyse the breakdown of proteins by hydrolysis of the peptide bond. Such enzymes tend to show specificity in that many will cleave only peptide bonds in which certain classes of amino acids form one of the partners. The enzyme trypsin will hydrolyse only those peptide bonds formed by the carboxyl groups of the long-chain basic amino acids lysine and arginine (Fig. 5.3.1, below). Chymotrypsin, on the other hand, rejects these peptide bonds and preferentially attacks those formed by the carboxyl groups of certain of the hydrophobic amino acids, notably the aromatic ones (Fig. 5.3.2).

Finally there may be a much lower degree of specificity. Enzymes exist which will cleave nearly all peptide bonds while others will hydrolyse nearly all esters, and so on.

It should also be noted that the specificity extends to the products. That is to say, in almost all cases a given enzyme will catalyse only one type of reaction with a particular substrate and will not assist in its conversion to any alternative products, even when these reactions seem to be chemically quite possible. This does not mean that an enzyme can never catalyse more than one type of reaction when given more than one type of substrate: e.g. proteolytic enzymes usually become esterases when confronted with peptides or amino-acid esters.

The potential of this great catalytic power coupled to great specificity has extremely important consequences. There is a very large number of reactions which it would be energetically possible

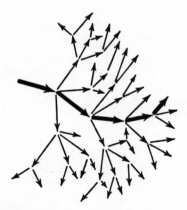

Fig. 5.1. Illustrating the power of a chain of enzymes to select just one of a large number of reaction pathways.

for a chemical compound (glucose, say) to undergo. The products of many of these reactions in their turn could each react in a large number of different ways, and so on for their products (Fig. 5.1). As will be seen below, enzymes cannot make energetically unfavourable reactions happen, they simply accelerate reactions that are possible. This acceleration is so great that it amounts to a *selection* process, and in the presence of suitable enzymes the starting material is steered through the maze of reactions as though the ever-multiplying possibilities did not exist. This power of enzymes to *organize* metabolism by virtue of the specificity and rapidity of action is even further increased by the way in which many of them are arranged in the cell, and by the many subtle mechanisms which provide for the increase or decrease of their activity as required (Chapter 23).

Before we can attempt to explain how enzymes function it is necessary to describe something of how chemical reactions take place. Consider the reaction

$$A + B \rightleftharpoons C + D$$

Even if the forward reaction is energetically favoured (p. 93) the following things must happen if the reaction is to proceed at a measurable rate. First A and B must collide in the proper orientation. (The majority of collisions in free solution are wasted since the permissible limits for the relative orientations of the particular bonds and groups in the reacting molecules fall within fairly narrow limits.) The second requirement is that the 'potential barrier' (Fig. 5.2) to the reaction must be overcome. Because of this barrier,

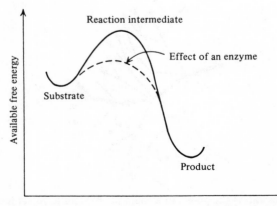

Fig. 5.2. A simplified view of the potential barrier in a chemical reaction.

many reactions do not take place even though they are energetically favoured. The barrier may be explained by the fact that while A and B are interacting to give C and D they form a temporary and unstable complex called a reaction intermediate. The free-energy state (see p. 93) of this reaction intermediate is often much higher than that of either products or reactants. In such cases the existence of an intermediate, though essential for the progress of the reaction, is at the same time a barrier. (In the absence of a catalyst the barrier can be overcome only by a favourable random fluctuation in the energy of the reactants. Heat will increase the magnitude of this fluctuation, which is why heat speeds up chemical reactions.)

Enzymes, like any other catalyst, are unable to promote a reaction unless it is energetically favourable. They cannot influence the initial or final energy states of the reactants and so cannot force an otherwise impossible reaction to occur. All they can do is act on both the factors (mentioned above) that control the rate of the reaction in such a way as to increase the speed with which a reaction, already possible on energetic grounds, will happen.

Those catalysts which act as surfaces, the class to which enzymes belong, can ensure the proper type of collision for reaction by binding A and B in the desirable specific orientation and in close proximity. Furthermore, if the binding is strong, A or B will remain on the enzyme until the arrival of its partner and there will be no need to wait for the relatively rare event of all three molecules, reactants and catalyst, colliding. (This is, in effect, increasing the local concentration of the reactants.) Thus the first thing an enzyme does is to bind its reactants. The usual forces that we have described before are employed, and specificity can be ensured by making use of the structural versatility of proteins described in Chapter 3. Fig. 5.3 shows how, in the two protein-degrading enzymes that we mentioned above, the amino-acid sequences differ so as to make sites with different binding properties for particular types of amino-acid side chain.

The analogy has frequently been made between the fitting of the correct substrate of an enzyme into the binding site designed for it and the fit between a key and a lock. This is a static picture and does not allow for the possibility that the enzyme might in some cases wrap itself around the substrate – in other words, that it will form the correct shape of lock only when the correct key is present. The picture of a lock and key is nonetheless helpful in understanding enzyme specificity, and you may feel that the idea stated in the preceding sentence is yet another instance of the much greater

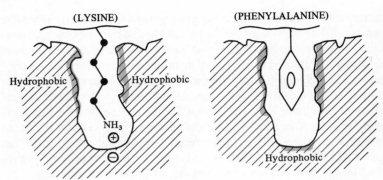

Fig. 5.3.1. The specificity
site of trypsin.

Fig. 5.3.2. The specificity site
of chymotrypsin.

sophistication of biological machines over most of those made by
man.

We can now explain the mode of action of so-called competitive
inhibitors. These are compounds which resemble the substrate
molecules sufficiently well to form some of the proper interactions
with the binding site but which are not sufficiently similar to take
part in the reaction and be released. The wrong key will not turn in
the lock, but if it is close enough to the right shape it may jam and
cannot then be removed. Thus, consider the enzymic conversion
of succinate ion to fumarate ion (Fig. 5.4 and see p. 137). When

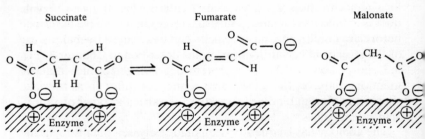

Fig. 5.4. Interactions of succinate dehydrogenase with the ionized forms of
succinic and malonic acids.

succinate is transformed to fumarate there is a marked change in the
geometry of the molecule and it can no longer be bound. Malonate
ion will bind to the appropriate site, since the spacing of its carboxyl
groups is not appreciably different from those of succinate, but
malonate cannot undergo the necessary change in order to be

released, and so it continues to occupy the binding site. While it remains, the enzyme is unable to accept any further molecules. Malonate is therefore a powerful inhibitor of the enzyme.

As an example of the utility of protein binding sites in raising the effective substrate concentration, we may consider the reactions that involve hydrolysis and in particular those in which protons are the active agent. The rate of such reactions will be dramatically enhanced if proton binding takes place. The concentration of protons at physiological pH is, of course, quite small: 10^{-7} mol/l at pH 7 by definition. (Incidentally, the volume of the cell is so small that this will represent only a hundred or so protons per cell.) The storage of a proton in a position in which it may be used in a hydrolytic reaction is equivalent to a drastic lowering of the pH at the site of storage. It does not, however, have all the undesirable consequences to the cell of a general increase in hydrogen ion concentration.

Students of thermodynamics will recognize that increasing the number of useful collisions represents an entropic contribution to the lowering of the free energy of the potential barrier. In addition enzymes may lower the potential barrier more directly. If some of the side chains on the protein surface are able to interact with the reaction intermediate they may stabilize it. The stabilization of a structure is synonymous with the lowering of its free-energy state. It can be shown that the probability of random thermal fluctuation sending the reactants over the potential barrier to form products increases exponentially with the lowering of the barrier; the reaction is accordingly speeded up in a dramatic way. Fig. 5.5 shows something of the proposed stabilization of a reaction intermediate in the case of the enzyme lysozyme, which hydrolyses polysaccharide chains in bacterial cell walls. We see in Fig. 5.5.2 the chain having already been cleaved by a proton donated from the carboxylic acid side chain marked 35. The resulting positively charged intermediate would be grossly unstable in free solution. It is given some stability here, first by the proximity of the negative charge of the carboxylic acid side chain marked 52 and secondly by interactions with the suitably shaped surface of the enzyme molecule which hold the ring in the deformed configuration necessitated by the geometry of the intermediate.

Fig. 5.5 also helps to illustrate one other feature of the relation between enzyme structure and function to which we have frequently referred. This is the placing of reactive groups in unusual environments which confer on them unusual properties. Glutamic

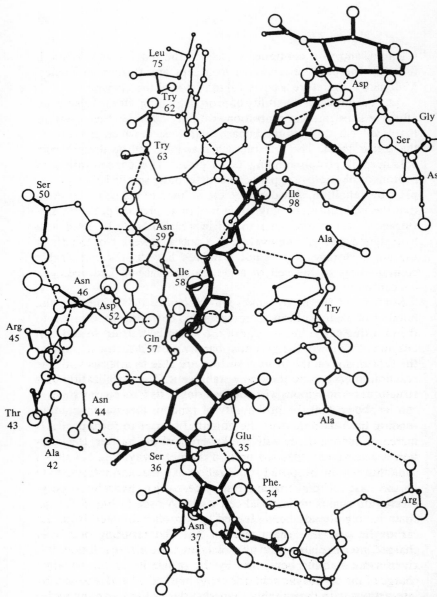

Fig. 5.5.1. A computer-drawn diagram showing the interaction between a hexasaccharide substrate (bold lines) and the amino-acid side chains of the binding site of lysozyme. Oxygen atoms are shown as large circles and nitrogen atoms by slightly smaller ones. Hydrogen atoms are omitted. Hydrogen bonds are represented by dotted lines. Note the complexity and specificity of the interactions. Some of the individual bonds are easily traced, e.g. the hydrogen bond between the CH_3CONH- substituent on the second sugar ring from the bottom and the side chain of asparagine 44. Others are less easy to follow without a stereoscopic drawing.

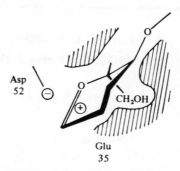

Asp
52

CH₂OH

Glu
35

Fig. 5.5.2. Idealized picture of the stabilization of the transitional intermediate formed during the hydrolysis of the bond between the second and
third sugars from the bottom in Fig. 5.5.1. Contacts with amino-acid side
chains stabilize the otherwise improbable conformation of the sugar ring.
The negative charge of Asp[52] stabilizes the positively charged intermediate
by means of an ionic interaction.

acid residue 35 is in an exceptionally hydrophobic environment. In
such an environment it is very much more difficult for the group to
ionize; in other words if it finds a proton, the proton will remain on
the glutamic acid residue until it is donated in the reaction. This then
is a very strong proton binding site. Aspartic acid residue 52, on the
other hand, is in a very hydrophilic environment, and under these
circumstances its ionization is favoured. Its environment therefore
helps it to remain charged, so that it is able to assist in stabilizing the
intermediate structure in the way referred to above.

We mentioned at the beginning of the chapter that the enzymes
were amenable to control, and that this feature greatly increased
their usefulness to the living organism. Control is exerted by a
variety of means, often involving a change in the structure of a
crucial region of the enzyme in response to the binding of a specific
molecule. Such phenomena are governed by exactly the same
general ideas as have been set out in this and the previous chapters.
Examples of the value of such control mechanisms are given in
Chapter 23.

Each organism probably contains many thousands of enzymes. It
seems likely that with the addition of co-factors and prosthetic
groups (which will be referred to frequently later on), all rely for
their remarkable properties on something like the same principles.
The enzymes give the chemical activities of the living state some of
the most striking of their unusual features.

6 Nucleic acids, polysaccharides and lipids

Nucleic acids

In the last chapter you will have formed the impression that the proteins have an important part to play in every distinctive activity of living processes save one. The apparent omission was reproduction. Even here proteins do have their role to play but the molecules

Table 6.1. *Bases and sugars in nucleic acids.* (*Note the various systems for numbering the atoms in the rings*)

Pyrimidines

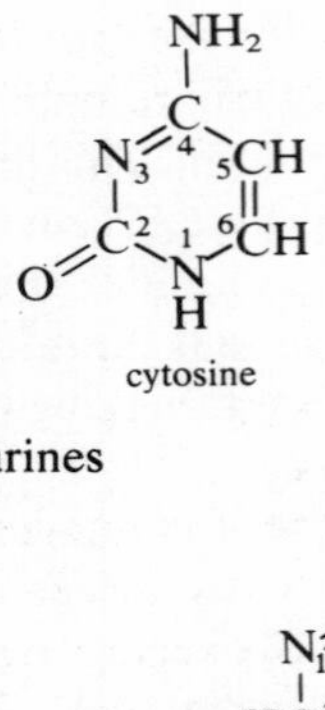

cytosine 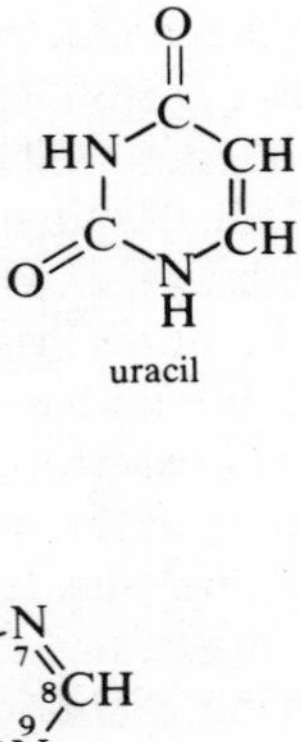   uracil 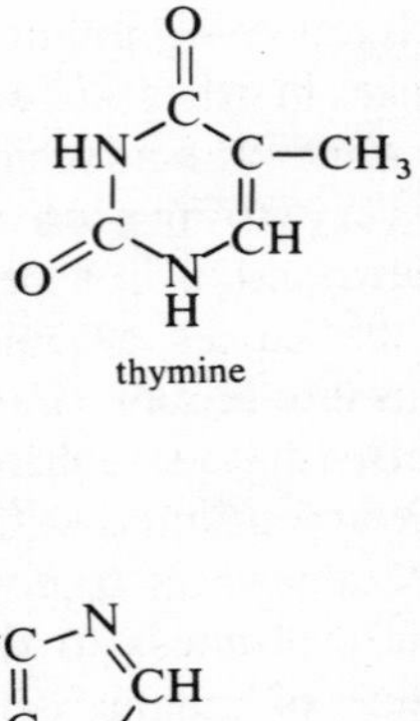   thymine

Purines

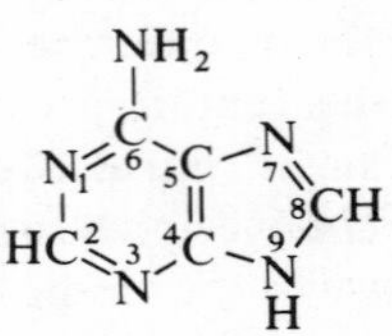

adenine 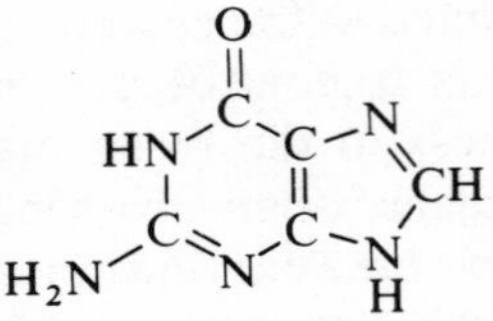   guanine

Sugars

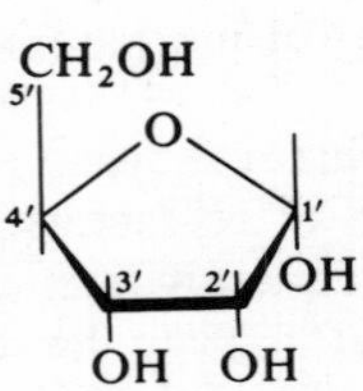

ribose 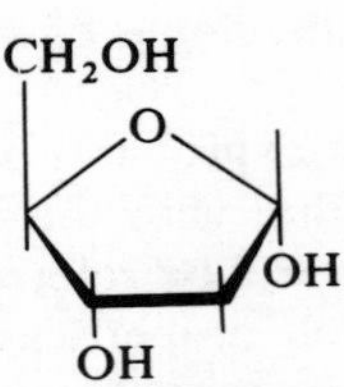   deoxyribose

most directly involved are the nucleic acids. This indeed is their only major function, and this fact explains why they possess a much narrower range of types of residue than the proteins. It also explains the relatively small number of types of nucleic acid.

The most common bases found in nucleic acids are shown in Table 6.1. They are linked by a glycosidic bond (see p. 7) to a sugar – D-ribose or D-aeoxyribose depending on whether the nucleic acid is ribonucleic acid (RNA) or deoxyribonucleic acid (DNA). The resulting base–sugar compound is called a *nucleoside*. When an –OH group in the sugar of a nucleoside is phosphorylated, the compound becomes a *nucleotide*. The nucleotides are linked in a specific order to form the nucleic acid by means of the phosphodiester linkage, as described previously.

Once again the full range of forces that we have described comes into play in the stabilization of the three-dimensional structure. The hydrogen bond receives most attention because of the remarkable way in which the bases, when in the same plane, fit into one another in two sets of complementary pairs stabilized by hydrogen bonds (Fig. 6.1). Fig. 6.1 shows how the bases lie in the same plane, and how they associate by means of two hydrogen bonds (in the case of A–T pairs) or three (in the case of G–C pairs). In no other pairing do the bases fit so well together, with the hydrogen bonds so readily established.

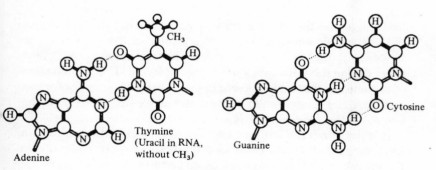

Fig. 6.1. Base pairing.

The famous double helical structure (Fig. 6.2) involves such co-planar pairs of bases. It will be seen that in each strand the bases are stacked upon each other like the steps in a spiral staircase. This structure is stable only when these pairs are complementary in the sense used above. Thus given the base sequence of one strand of DNA, there is only one possible sequence for the opposing strand

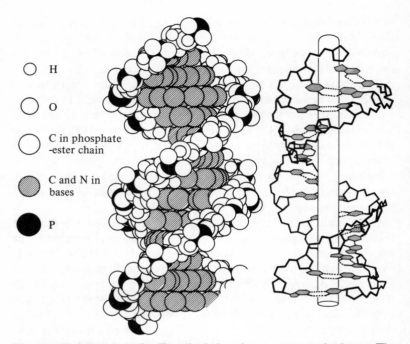

H

O

C in phosphate
-ester chain

C and N in
bases

P

Fig. 6.2. The double helix. The shaded regions represent the bases. The
sugar rings are unshaded.

that will permit the two to form a stable double helix. This fact is of
the deepest significance. We shall see later in the book that the
sequence of one strand *is* used to predetermine the sequence of the
other during the self-replication of the molecule. This specification
of sequence is the primary event in the transfer of genetic information from one generation to the next and (as we shall also see) in the
control of the sequences of RNA and protein molecules.

It is not generally realized that the hydrogen bonds add only a
final, though critical, measure of stability to a structure already on
the verge of being held together by much more powerful forces. Of
the non-covalent forces previously mentioned, hydrophobic and
dipole interactions stabilize the stacking of the bases in each strand
and also push the two strands together. It is in fact this squeezing
action that forces water molecules out of the centre of the assembly
and leaves a clear field for the hydrogen bonds to form (Fig. 6.3).

The double helical structure is characteristic of DNA. A closely
analogous double helix can exist for RNA wherever complementary base pairs are found. This structure is somewhat different from

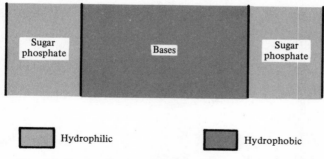

Fig. 6.3. Hydrophobic and hydrophilic regions in DNA.

double-stranded DNA because of the intrusion of the bulk of the extra oxygen atom at position 2′ of the ribose (Fig. 2.2) into a crowded part of the structure. Fig. 6.4 shows the structure in outline. The bulk of the 2′ –OH is sufficient to ensure that this structure will also be that adopted by double helices in which one strand is DNA and the other is RNA. Such *DNA–RNA hybrids* are of great importance in the initial stages of synthesis of proteins (Chapter 19), and it might well be useful for purposes of recognition to have such a marked difference in overall conformation between the DNA–RNA hybrid and double-stranded DNA, since the latter has a quite distinct role to play.*

The double helical structure of RNA is most frequently found only in stretches of the molecule where complementary sequences happened to occur in positions that enable them to come together.

Fig. 6.4. Double-stranded RNA. In contrast to DNA, the bases are neither parallel to one another nor perpendicular to the axis of the helix.

* Double-stranded DNA can adopt the RNA-type helix, but since it lacks the 2′ –OH it is not forced to: there is reason to believe that in the cell it is normally found in the characteristic configuration shown in Fig. 6.2.

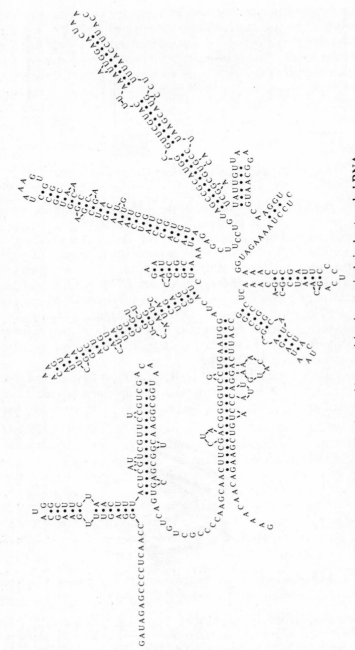

Fig. 6.5. Regions of base pairing in a typical single-stranded RNA.

These double-stranded regions are separated by single-stranded regions where no complementary sequence is available (e.g. Fig. 6.5). Such a structure will be in strong agitation through random molecular bombardment and should not be thought of as in any way a static structure. Divalent cations such as Mg^{2+} will help to make the various parts of the structure cohere, by acting as a positively charged adhesive between two negatively charged phosphate groups on different parts of the structure. No other types of cross-linking are known, in contrast with the proteins and polysaccharides.

DNA is found outside cells as single- or double-stranded virus nucleic acid 10^3–10^5 bases in length. Inside the cell most of the DNA is found in the nucleus in the form of chromosomes, which are complexes between double helical DNA, proteins (of which many are basic) and some lipids. This DNA, as we have already hinted, acts as the library of genetic information for the cell and this aspect will be dealt with more fully in Chapter 18. It has a very large chain length, hundreds of thousands of bases, and a relative molecular mass in the hundreds of millions. The DNA of a typical bacterium is 1.4 mm long, and were it not for its thinness it would be visible to the naked eye. Remarkably, in spite of the technical difficulties, methods have been developed for the determination of the sequences in DNA, and Fig. 6.6 shows a section of DNA that encodes genetic information in a virus.

RNA is found outside cells as virus nucleic acid. It is most usually single stranded and is 10^3–10^4 bases in length. Inside the cell there are three types, all of which function in protein synthesis. *Ribosomal* RNA (rRNA) occurs in ribosomes, which are the subcellular components involved in protein synthesis. There are three nucleic-acid components in the ribosome. One is of about 3×10^3 bases, one is half that length and one is just over 10^2 bases. Any double-stranded structure in them is of the type discussed on p. 55. *Messenger* RNA (mRNA) carries the instructions for assembling amino acids from the nucleus to the ribosome. Messenger RNA is a single-stranded molecule of a length of the order of 10^3–10^4 bases. Fig. 6.5 represented, in fact, a portion of a mRNA.

Transfer RNAs (tRNA) carry the amino-acid residues to the ribosome, to be incorporated into protein in an order dictated by the hydrogen-bonding properties of the nucleotide sequence of the messenger (Ch. 20). Each amino acid has at least one species of tRNA molecule of its own, with about 70–80 bases in a characteristic sequence that is unique to that tRNA species. Transfer RNA

GTATTGCTTCTGCTCTTGCTGGTGGCGCCATGTCTAAATTGTTTGGAGGCGGTCAAAAAGCCGCCTCCGGTGGCATTCAAGGTGATGTGC

TTGCTACCGATAACAATACTGTAGGCATGGGTGATGCTGGTATTAAATCTGCCATTCAAGGCTCAAGGCTCTAATGTTCCTAACCCTGATGAGGCCG

CCCCTAGTTTGTTTCTGGTGTGCTATGGCTAAAGCTGGTAAAGCTTCTTGAÂGGCTACGTTGCAGGCTGGCACTTCTGCCGTTTCTGATA

AGTTGCTTGATTTGGTTGGACTTGGTGGCCAAGTCTGCCGCTGATAAAGGAAAGGATACTCGTGATTATCTTGCTGCTGCATTCCTGAGC

TTAATGCTTGGGAGCGTGCTGGTGTGCTGATGCTTCCTCGCTGGTATGGTTGACCCCGATTTGAGAATCAAAAAGAGCTTACTAAAATGC

AACTGGACAATCAGAAAGAGATTGCCGAGATGCAAAATGAGACTCAAAAAGAGATTGCTGGCATTCAGTCGGCGACTTCACGCCAGAATA

CGAAAGACCAGGTATATGCCACAAAATGAGATGCTTGCTTATCAACAGAAGAAGAGTCTACTGCTCGCGTTGCGTCTATTATGGAAAACACCA

ATCTTTCCAAGCAACAGCAGGTTTCCGAGATTATGCGCCAAATGCTTACTCAAGCTCAAACGGCTGGTCAGTATTTTACCAATGACCAAA

TCAAAGAAATGACTCGCAAGGTTAGTGCTGAGGTTGACTTAGTTCATCAGCAAACGCAGAATCAGCGGTATGGCTCTTCTCATATTGGCG

CTACTGCAAAGGATATTTCTAATGTCGCTAATGTCTGCTTCACTGATGCGTCCTTCTGTGGTTGATATTTTTCATGCTGTGGTTGCCGATA

Fig. 6.6. A portion of the base sequence of the DNA of a virus. The part shown is about one-sixth of the total sequence, and contains instructions for the synthesis of part of one of the virus proteins.

59

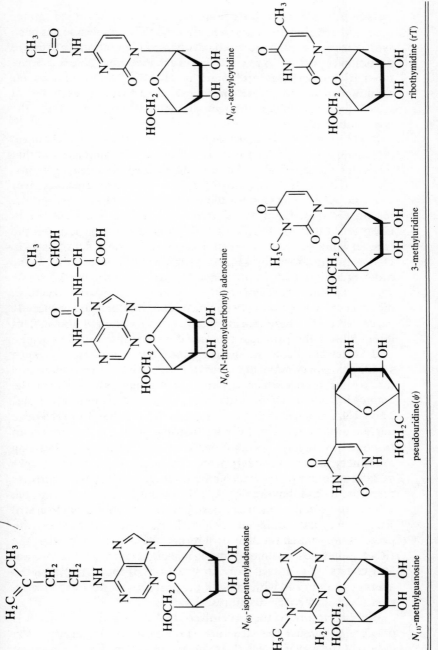

The bases are shown as nucleosides.

molecules are notable for containing a range of so-called 'minor bases'. These are bases on which, after the tRNA molecule has been synthesized, the cell has carried out chemical operations to produce a modified structure. A few such bases are shown in Table 6.2. Some types of minor bases are found in all tRNA molecules so far examined, while others are considerably less common: each type of tRNA molecule makes its own characteristic selection from the possibilities available.

It will be noticed from Table 6.2 that the effect of the chemical modifications is to widen the range of non-covalent interactions that the bases can undertake. We see enhanced possibilities for hydrophobic and for all types of hydrophilic interactions. This fact gives us a clue as to the reason for the existence of minor bases, and the reason for their being found mainly in tRNA. We shall see in Chapter 20 that the function of tRNA is to mediate between the form in which genetic information is stored (as the order of bases in DNA) and the form in which that information is expressed (the order of amino acids in proteins). Nucleic acids are on the whole very long molecules of fairly uniform secondary structure: differences between one sequence and another are generally perceived by the base-pairing interactions mentioned above. This interaction is the principal non-covalent one involved in nucleic-acid structure. Proteins, on the other hand, are usually much smaller, compact molecules, with a wide variety of secondary and tertiary structures which depend (as we have stressed) on the amino-acid sequence via a full range of non-covalent interactions. The tRNA molecules must be able on the one hand to recognize nucleic-acid sequences by base pairing, and, on the other, to respond to highly specific parts of the protein-synthesizing machinery, which are largely protein in nature and require all types of interactions for their recognition. It should not be a surprise therefore to find that the tRNA molecule, in addition to being able to undertake complementary base pairing with other nucleic-acid strands, has also equipped itself with the means to interact with proteins very much on their own terms. We may note too that the tRNA molecules are unique among nucleic acids in that they are not long chains but compact, globular structures of about the size of the average protein and with some of the proteins' tendency to unique, characteristic tertiary structures.

We shall return to the structure of tRNA on p. 204, where it is possible to discuss the structure more explicitly in terms of the details of the way in which the molecule functions. For the moment you should glance briefly at Fig. 20.1, p. 205, and note the alternation of the double- and single-stranded regions, which were referred to above.

Nucleic acids are frequently found in combination with basic proteins (histones, nucleoproteins, ribosomal proteins, etc.). The precise mode of action of these proteins is obscure, but they may contribute an essential element to the proper structure of the final nucleoprotein assembly. Ionic forces probably mediate in the interactions between nucleic acid and basic proteins. Some of the basic proteins found in conjunction with nucleic acid have as much as 80% of their side chains drawn from the basic group in Table 3.1. Fig. 6.7 shows the amino-acid sequence of a histone which, while it does not have quite such an extreme proportion of basic residues, still has a great many more than the average protein. Many ionic bonds may form between the positively charged side chains of such proteins and the negatively charged phosphate groups of the sugar-phosphate backbone.

Acetyl - Ser - Gly - **Arg** - Gly - **Lys** - Gly - Gly - **Lys** - Gly - Leu - Gly - **Lys** - Gly - Gly - Ala
10

Lys - **Arg** - **His** - **Arg** - **Lys** - Val - Leu - **Arg** - Asp - Asn - Ile - Gln - Gly - Ile - Thr
20   30

Lys - Pro - Ala - Ile - **Arg** - **Arg** - Leu - Ala - **Arg** - Gly - Gly - Val - **Lys** - **Arg** - **Arg**
40

Ile - Ser - Gly - Leu - Ile - Tyr - Glu - Glu - Thr - **Arg** - Gly - Val - Leu - **Lys** - Val
50   60

Phe - Leu - Glu - Asn - Val - Ile - **Arg** - Asp - Ala - Val - Thr - Tyr - Thr - Glu - **His**
70

Ala - **Lys** - **Arg** - **Lys** - Thr - Val - Thr - Ala - Met - Asp - Val - Val - Tyr - Ala - Leu
80   90

Lys - **Arg** - Gln - Gly - **Arg** - Thr - Leu - Tyr - Gly - Phe - Gly - Gly
100

Fig. 6.7. The amino-acid sequence of a histone. The basic residues are emphasized.

Some other proteins found in association with nucleic acid do not rely so much on ionic bonds. These are the virus proteins which are so constructed as to have a strong quaternary structure. This encloses and protects the nucleic acid (Fig. 6.8).

Polysaccharides

We saw in Chapter 2 how sugar molecules can form chain polymers by means of the glycosidic link. But in order to appreciate the full range of possibilities available to living matter in the design of polysaccharides, we must look a little more closely at the properties of the sugar monomers, or *monosaccharides*. Originally, the chemical behaviour of the monosaccharides suggested that they consisted

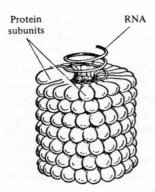

Fig. 6.8. Quaternary structure in a cylindrical virus.

of straight carbon chains with –OH groups on each carbon except one, which carried either an aldehyde or a keto function. However, it is well known to organic chemists that aldehyde and keto groups can react with $>$C–OH groups. A further examination of the chemical properties of the monosaccharides produced evidence that this reaction is strongly favoured, with the result that the straight-chain form gives rise to a ring (Fig. 6.9). It will be observed

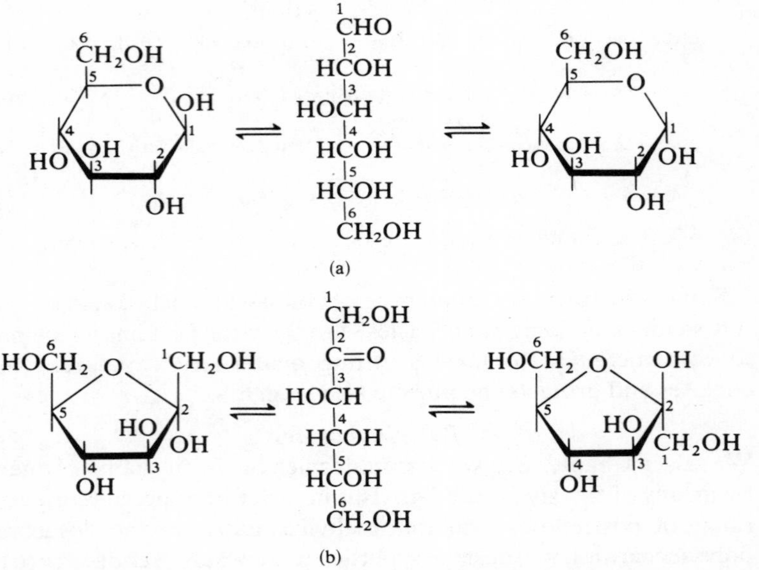

Fig. 6.9. Straight-chain and ring formulae for (a) glucose, an aldo-sugar and (b) fructose, a keto-sugar.

that the ring closure can take two slightly different paths and give rise to two slightly different products, which are normally designated the α and β forms of the sugar. These two forms, which differ in the position with respect to the ring of the –OH group on C-1, are readily interconvertible via the straight-chain form. A solution of a monosaccharide will normally consist of a mixture of the two forms.

We saw in Chapter 2 that the glycosidic bond is formed between the –OH on C-1 and any of the –OH groups on the next sugar (except the one that is already involved in forming the ring of that sugar). Thus in the case of polymers of glucose, the glycosidic bond can be formed between C-1 of one molecule and the –OH group on C-1, C-2, C-3, C-4 or C-6 of the next. The bond would be designated 1→1, 1→2, 1→3, 1→4 or 1→6, respectively. If the –OH group on C-1 had been in the α position, the bond would be designated 1→1α, 1→2α, and so on, while if it had been in the β position the bond would be designated 1→1β, 1→2β, etc. There are thus a great many possibilities: ten possible disaccharides of glucose as compared to one possible dipeptide of a single amino acid. Having looked at the structures of other biological polymers, we should not be surprised to learn that the biosynthetic machinery is sufficiently discriminating as to be able to avail itself only of the particular form of glycosidic bond required in any instance. What is perhaps more surprising is the fact that such apparently trivial structural differences can give rise to remarkable differences in functional properties: we shall see in a moment that the structure in Fig. 6.10 is found in cellulose and can produce objects of great mechanical strength, while the rather similar-looking chain in Fig. 6.11 is found in starch and confers no mechanical strength at all.

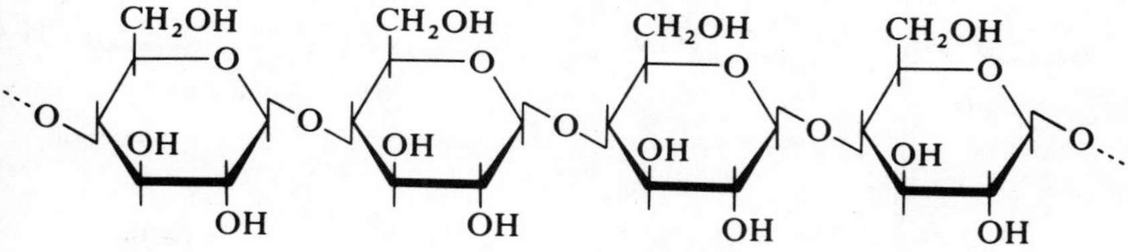

Fig. 6.10. A portion of a 1 → 4β linked chain of glucose residues.

A number of six- and five-carbon sugars form polymers (Table 6.3), although several important polysaccharides consist of glucose alone. Even when more than one type of monosaccharide is found in a polysaccharide, we do not see the complicated, rigorously controlled sequences that are characteristic of proteins and nucleic

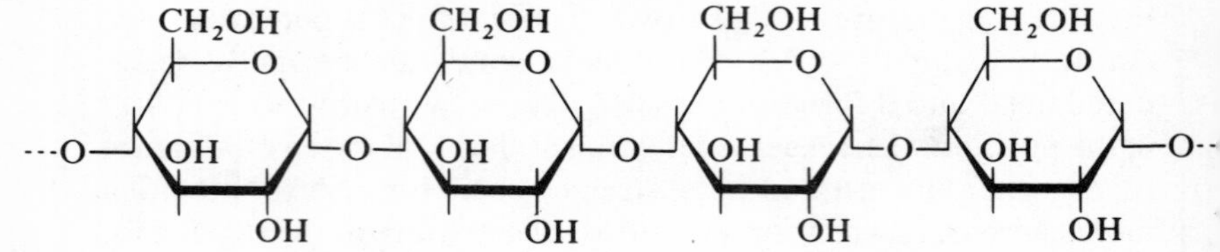

Fig. 6.11. A portion of a $1 \to 4\alpha$ linked chain of glucose residues.

acids. In some instances the sequence is controlled to the extent of ensuring the strict alternation of two types of monosaccharide (i.e. –A–B–A–B–A–B–), and in other instances control is so lax that a sample of a given polysaccharide will consist of a mixture of molecules in which several different sequences are represented.

Table 6.3. *Some sugar molecules found in polysaccharides*

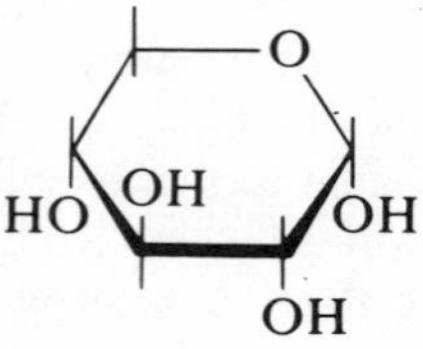

D-xylose 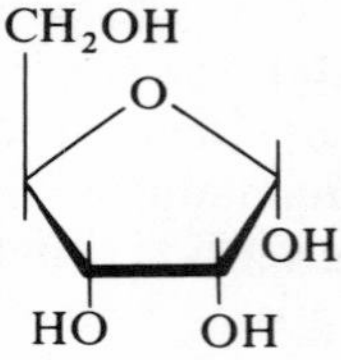 D-ribose 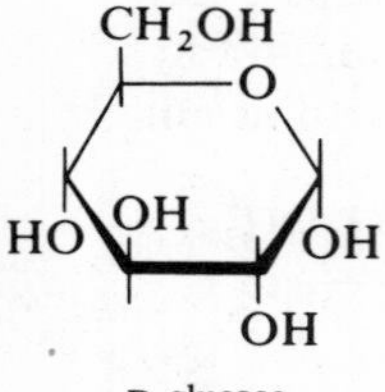 D-glucose 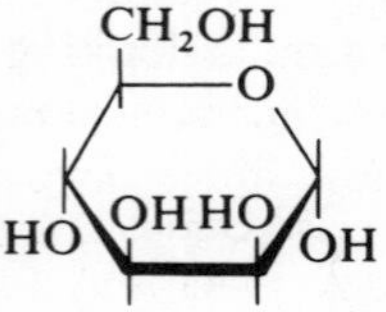 D-mannose 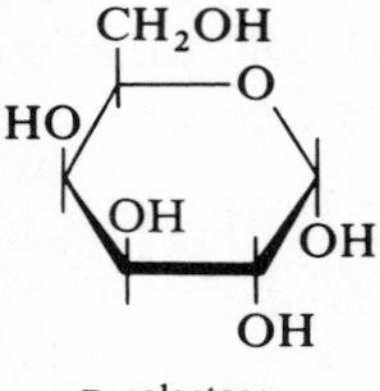 D-galactose 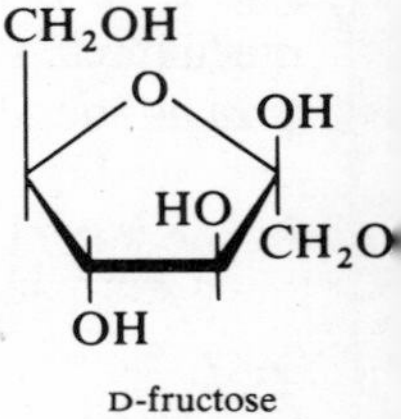 D-fructose 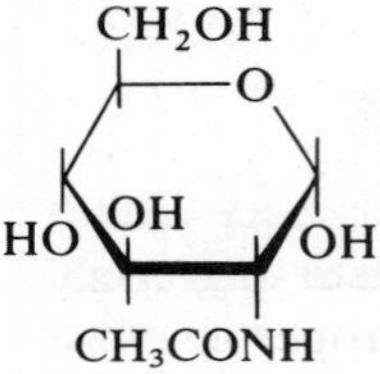 N-acetyl-D-glucosamine 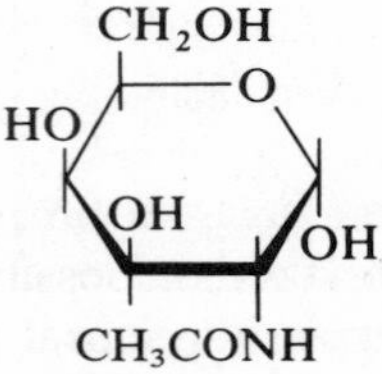 N-acetyl-D-galactosamine 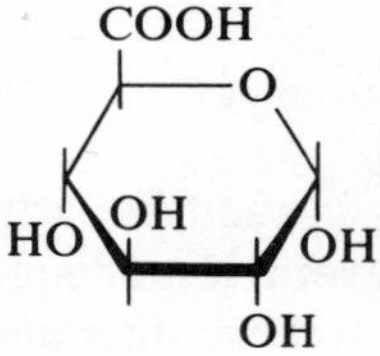 D-glucuronic acid 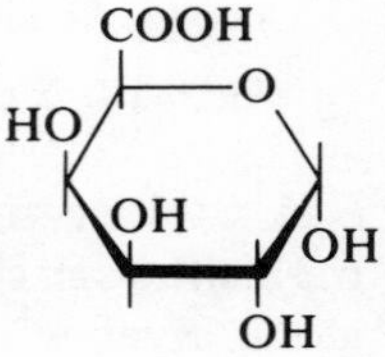 D-galacturonic acid

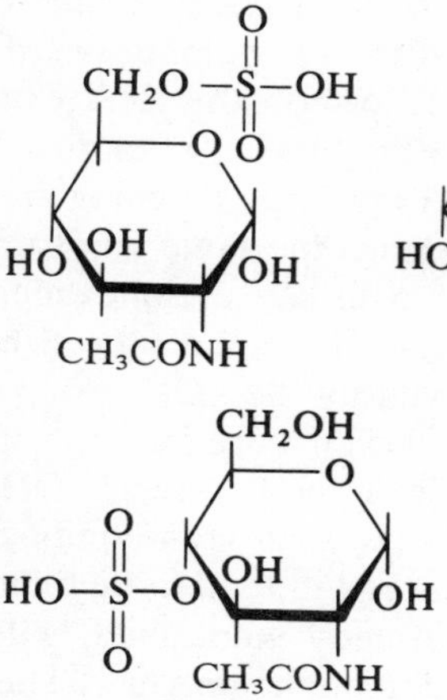

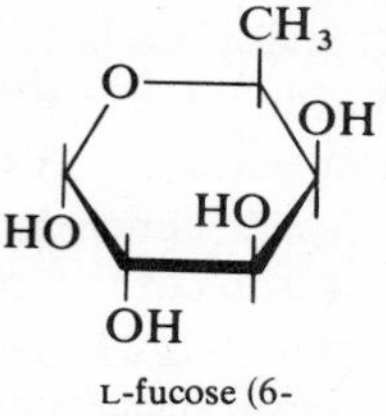

L-fucose (6-deoxy-L-galactose)

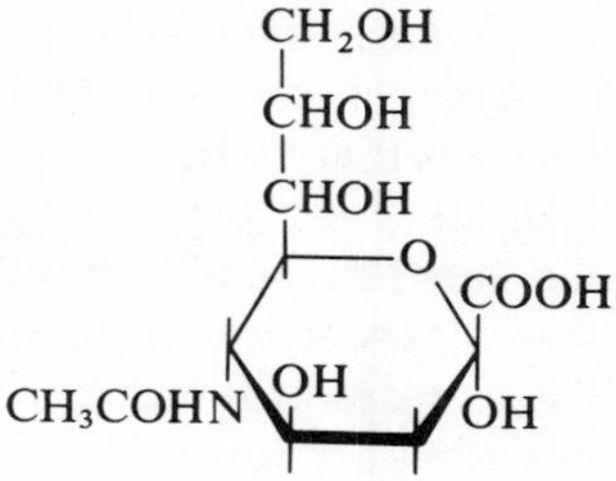

N-acetyl-D-neuraminic acid

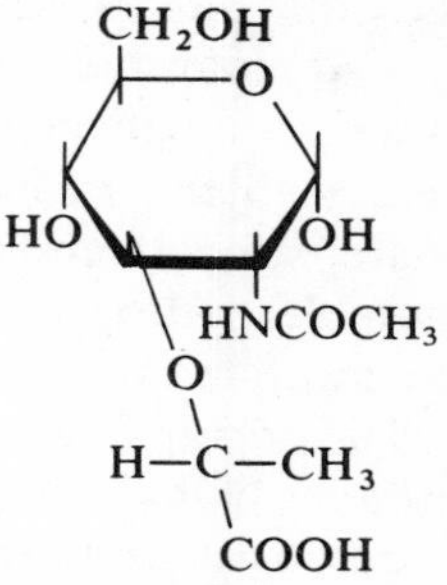

N-acetyl-D-glucosamine sulphates

N-acetyl-D-muramic acid

We have said that the availability of a variety of types of glycosidic bond permits a surprisingly wide range of properties. These possibilities can be further augmented by the cell's carrying out chemical modifications to the monosaccharides. This is reminiscent of the augmentation of the properties of mononucleotides that we have already discussed, but in the present case the modifications normally occur before the monomer units are incorporated into polymers. The range of possible alterations is quite wide, and modifications exist that confer the ability to participate in all the kinds of non-covalent interaction. The latter part of Table 6.3 shows some particularly common examples.

Polysaccharides may be of considerable size, although, in distinction to proteins and nucleic acids, they often exist as mixtures of molecules of different chain lengths. It is then possible to speak only of an *average* value for this and other structural properties. The three-dimensional configuration is controlled by the usual forces, including, surprisingly enough for molecules in which the monomer unit has so many –OH groups, the hydrophobic interaction. For instance, in the most stable configuration for the D-glucose ring it happens that all the –OH groups are turned so as to point outward from the edges of the ring. (It is not possible to show this clearly in the illustrations of polysaccharide structures in this book: to do so would obscure other important features.) A molecule touching the flat surfaces of the ring would experience much the same kinds of interaction as it would do if it was in contact with, say, cyclohexane, a strongly hydrophobic molecule. This having been said, the rich possibilities for hydrogen bonding preserved by the –OH groups should not be forgotten, both for glucose and even more for some of the other sugars, which cannot so easily accommodate their –OH groups along the edge of the rings. When the hydrogen bonding takes place between sugar rings the structures will tend to be tight and, with the hydrophobic interactions already mentioned, will confer strength and a low degree of hydration on the polymer. The more the hydrogen bonding tends to be made to molecules of water the more the structure tends to be soft, open and highly hydrated.

The principal functions of the polysaccharides mirror those of the proteins that we discussed in the last chapter – they can act as food stores, or as structural components, and (it is becoming clear) some are able to behave like the other class of proteins discussed in that chapter, by undertaking specific binding interactions.

Both animal and plant tissues make use of polysaccharides as food storage materials. We shall see in the next section of this book that monosaccharides are a valuable source of metabolic energy. However, they would be difficult to store in the monomeric form because they would tend to diffuse rapidly away from the site of storage, and if concentrated in a relatively small volume would produce a solution of inconveniently high osmotic pressure. Both of these problems are solved if the monosaccharides are incorporated into polymers. The rate of diffusion falls to a very low value, and the osmotic pressure exerted by a given mass of sugar is vastly less when it is in the form of a large polymer than when it is in the form of the free monomer. Glycogen typifies the animal storage polymer, starch the plant product. Glycogen consists of chains of glucose molecules joined by glycosidic links in the $1 \rightarrow 4\alpha$ configuration and cross-linked by glycosidic links chiefly in the $1 \rightarrow 6\alpha$ configuration (Fig. 6.12). There are approximately 10^4 glucose residues per molecule. There is only approximate control over the lengths of chains and degree of cross-linking, and so the molecules form a heterogeneous population.

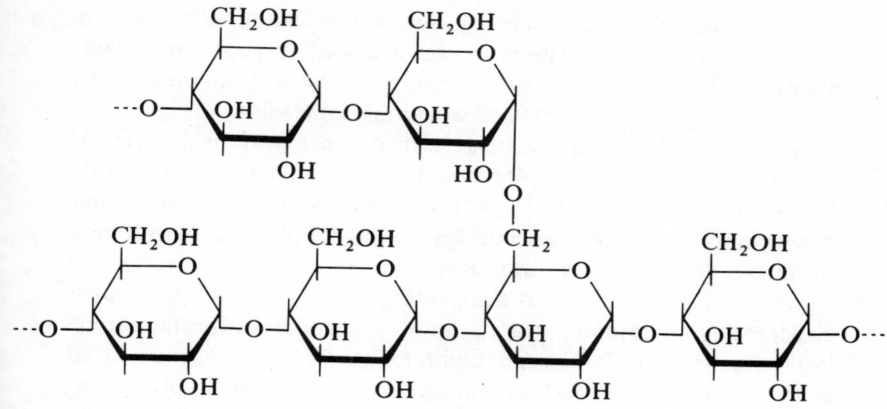

Fig. 6.12. A portion of the glycogen molecule.

Starch consists of two types of molecule. One is known as amylopectin and the other as amylose. Amylopectin has a structure similar to glycogen but contains on average rather fewer glucose residues. Amylose is a straight-chain polysaccharide containing 10^2–10^3 glucose residues joined by glycosidic linkages in the $1 \rightarrow 4\alpha$ configuration (Fig. 6.11). (The indeterminacy of structure in food storage polysaccharides is not paralleled by any indeterminacy in food storage proteins, which appear to have sequences as rigidly defined as any other protein. This difference is probably because certain quite small divergences from the correct structure could render such proteins useless (for example by their being made insoluble) in a way which is not easy to imagine for polysaccharides.)

Cellulose is the best known example of a polysaccharide with considerable mechanical strength. It is formed of unbranched chains of glucose residues joined by the glycosidic linkage in the $1 \rightarrow 4\beta$ configuration (see Fig. 6.10). Such chains can pack closely together and form a strong bundle with many possibilities for hydrogen bonding between them. For greater strength, in wood for example, extensive cross-linking by aromatic free-radical reactions may occur, in a way somewhat reminiscent of the proteins (p. 33). Many other sugar polymers are known with similar functions to those of cellulose; all show extensive non-covalent bonding between sugar rings.

Not all structural polysaccharides rely on strength or rigidity to fulfil their functions. Some, particularly those with appropriately modified monosaccharides, have hydration properties that cause them to form solutions with great value as lubricants, for example in the joints of animals. These solutions are often superior to conventional lubricants used in machinery in the way in which their viscosity varies with the rate of working of the joint.

Storage and structural polysaccharides are widely distributed among micro-organisms, plants and animals. They are frequently found in combination with proteins (where they are called glycoproteins or mucopolysaccharides) and with lipids. Such combined materials have many important structural and mechanical functions, although it is, at the present state of knowledge, much harder to give explanations for their actions in molecular terms. Nonetheless, if you are familiar with the A, B, O blood groups used in the matching of blood for transfusion, you may be interested to look at Fig. 6.13. This Figure indicates in outline the crucial structural feature of the cell-surface glycoprotein that determines to which group an individual belongs. If blood is mis-matched, the different shapes of the determinant features may interact with antibodies with appropriately tailored binding sites (cf. Fig. 4.4) with serious consequences for the recipient. This is a rather artificial example of the specific binding properties of a polysaccharide chain, since it deals not with its normal function, but with what happens when things go wrong as a result of ill-judged human intervention. It would have been more satisfying to have shown such molecules as they are meant normally to act, for example in specific interactions between the surfaces of two cells that are in contact. Such interactions are likely to prove to be of the greatest importance in, for example, the intercommunication between cells during the growth and differentiation of tissues: the potential rewards for explaining them in molecular terms are almost certain to be immense.

Now that we have completed our review of the polysaccharides, we will reiterate what we said at the start. The principles that link the structure of the polysaccharides to their function are certainly those that apply in the case of the other macromolecules. The more rudimentary state of our knowledge of these molecules (although it is not much more rudimentary than was that of proteins and nucleic acids twenty-five years ago) makes it more difficult to give detailed explanations of individual pieces of behaviour. Nonetheless our knowledge is improving; the polysaccharides are so important, particularly to the developing science of the study of the interrelationship between cells, that we may expect our understanding of their structure and function to grow rapidly in the next few years.

69

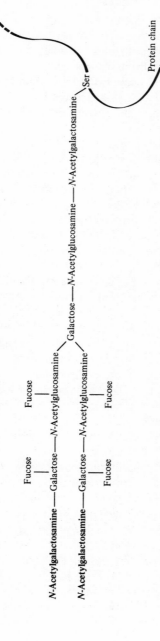

Fig. 6.13. The structural features responsible for the A, B, O blood groups. The figure shows a branched polysaccharide chain, covalently attached to the serine residue of a protein chain. The resulting glycoprotein is abundant on the surface of red blood cells. The structure shown is that found on cells of group A. In group B the two sugars that are printed in bold type are replaced by N-acetylglucosamine residues. In cells of group O they are missing.

Lipids

We have seen that proteins, nucleic acids and polysaccharides are large molecules produced by covalent polymerization of small monomer units. Single lipid molecules are on average a little larger than the monomers of these three classes, but substantially smaller than the macromolecules. However, the individual lipids associate non-covalently to form very large structures indeed, and it is this that justifies their inclusion in the section of this book devoted to macromolecules.

There now follows a brief catalogue of the major types of lipid. Without some guiding theme it would be both tedious and rather pointless to read through it. But fortunately there is such a theme, and a very important one. We suggest, therefore, that you glance only briefly at Table 6.4, which lists some of the more important types of lipid, and return to it after you have read the next few pages.

The key to the understanding of the lipids is that, while building units for the proteins, nucleic acids and polysaccharides are relatively limited in number, the lipid molecules that are analogous to these building units are very numerous, being themselves assembled by permutation of many *still smaller* units. There are thus many more types of completed molecule, the varied characteristics of which are *again* permuted when these molecules associate to form larger structures. The result is a truly immense range of ways in which the possibilities of non-covalent interactions can be exploited. Furthermore, the completed lipid molecules are brought together in the larger structures, not by a covalent linkage like the peptide, phosphodiester, or glycosidic bonds, but by non-covalent, largely hydrophobic, interaction.

The large lipid assemblies can thus, in a sense, be seen as macromolecules that have freed themselves even more than the other macromolecules from the limitations of the covalent bond. As the other classes used non-covalent interactions to transcend the limitations of small molecules, many of the properties of which are dominated by the inflexible characteristics of the covalent bond, so the lipids might be expected to be capable of even more subtlety and flexibility in their behaviour. We shall indeed see later that the lipids are involved, along with other types of molecule, in some quite remarkable activities, particularly those involving the membranes of cells. Although few of these phenomena have yet received a completely satisfactory explanation, an understanding of them must be sought at least in part in the properties of the lipids.

Table 6.4. *Some of the major types of lipid molecule*

a. *Complex lipids**

```
                                    CH₃
                                    |
                                    CH₂
                                    |
                                    CH₂
                                    |
                                    CH₂
                                    |
                                    CH₂
                                    |
                                    CH₂
                                    |
                                    CH₂
                                    |
                                    CH₂
                                    |
                                    CH₂
                                    |
                                    CH₂
                        CH₃         CH₂
                        |           |
                        CH₂         CH₂
                        |           |
                        CH₂         CH₂
                        |           |
                        CH₂         CH₂
                        |           |
                        CH₂         CH₂      (alcohol residue)
                        |           |
                        CH₂         CH₂
                        |           |
                        CH₂         CH₂
                        |           |
                        CH₂         CH₂
        (fatty-acid residue)  CH₂   CH₂
                        |           |
                        CH₂         CH₂
                        |           |
                        CH₂         CH₂
                        |           |
                        CH₂         CH₂
                        |           |
                        CH₂         CH₂
                        |           |
                        CH₂         CH₂
                        |           |
                        CH₂         CH₂
                    O=C- O - CH₂
```

A component of
beeswax

* The particular importance of those parts of the structures that are shown in bold type is discussed on p. 80 and p. 81.

Table 6.4—*continued*
a. Complex lipids—*continued*

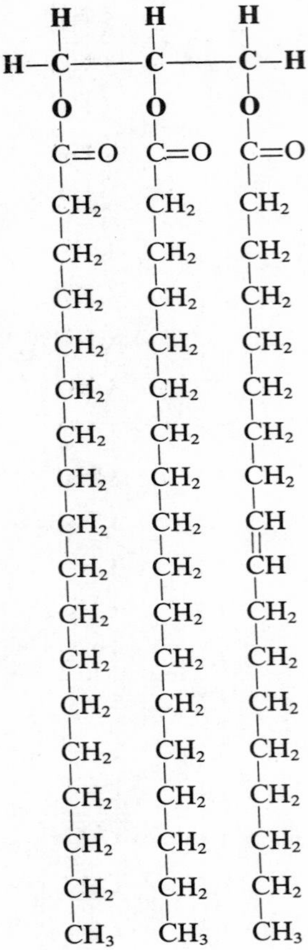

$$
\begin{array}{ccc}
\text{H} & \text{H} & \text{H} \\
| & | & | \\
\text{H-C} & \!\!\!\!\text{-C-} & \!\!\!\!\text{C-H} \\
| & | & | \\
\text{O} & \text{O} & \text{O} \\
| & | & | \\
\text{C=O} & \text{C=O} & \text{C=O} \\
| & | & | \\
\text{CH}_2 & \text{CH}_2 & \text{CH}_2 \\
| & | & | \\
\text{CH}_2 & \text{CH}_2 & \text{CH}_2 \\
| & | & | \\
\text{CH}_2 & \text{CH}_2 & \text{CH}_2 \\
| & | & | \\
\text{CH}_2 & \text{CH}_2 & \text{CH}_2 \\
| & | & | \\
\text{CH}_2 & \text{CH}_2 & \text{CH}_2 \\
| & | & | \\
\text{CH}_2 & \text{CH}_2 & \text{CH}_2 \\
| & | & | \\
\text{CH}_2 & \text{CH}_2 & \text{CH}_2 \\
| & | & | \\
\text{CH}_2 & \text{CH}_2 & \text{CH} \\
| & | & \| \\
\text{CH}_2 & \text{CH}_2 & \text{CH} \\
| & | & | \\
\text{CH}_2 & \text{CH}_2 & \text{CH}_2 \\
| & | & | \\
\text{CH}_2 & \text{CH}_2 & \text{CH}_2 \\
| & | & | \\
\text{CH}_2 & \text{CH}_2 & \text{CH}_2 \\
| & | & | \\
\text{CH}_2 & \text{CH}_2 & \text{CH}_2 \\
| & | & | \\
\text{CH}_2 & \text{CH}_2 & \text{CH}_2 \\
| & | & | \\
\text{CH}_2 & \text{CH}_2 & \text{CH}_2 \\
| & | & | \\
\text{CH}_2 & \text{CH}_2 & \text{CH}_2 \\
| & | & | \\
\text{CH}_3 & \text{CH}_3 & \text{CH}_3
\end{array}
$$

a fat

Table 6.4—*continued*
a. Complex lipids—*continued*

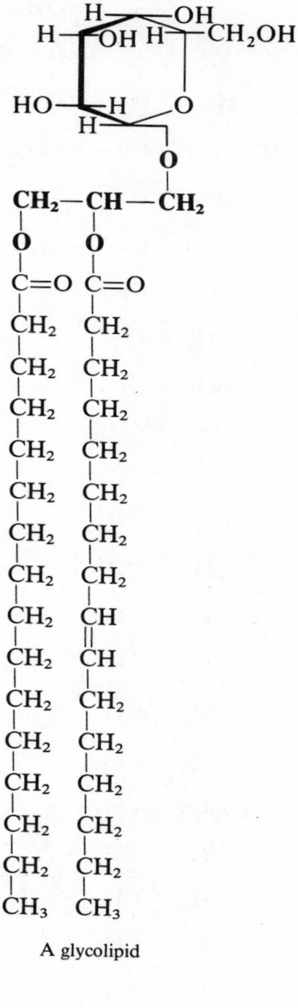

A glycolipid

Table 6.4—*continued*
a. Complex lipids—*continued*

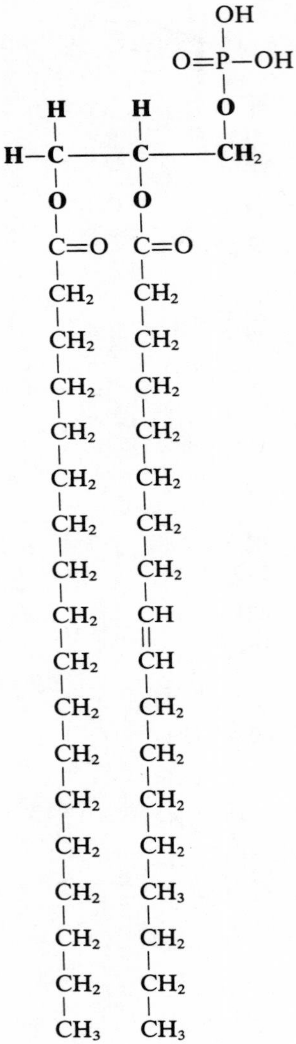

```
                        OH
                        |
                    O=P—OH
     H      H        |
     |      |        O
 H—C——————C——————CH₂
     |      |
     O      O
     |      |
     C=O    C=O
     |      |
     CH₂    CH₂
     |      |
     CH₂    CH₂
     |      |
     CH₂    CH₂
     |      |
     CH₂    CH₂
     |      |
     CH₂    CH₂
     |      |
     CH₂    CH₂
     |      |
     CH₂    CH₂
     |      |
     CH₂    CH
     |      ‖
     CH₂    CH
     |      |
     CH₂    CH₂
     |      |
     CH₂    CH₂
     |      |
     CH₂    CH₂
     |      |
     CH₂    CH₂
     |      |
     CH₂    CH₃
     |      |
     CH₂    CH₂
     |      |
     CH₂    CH₂
     |      |
     CH₃    CH₃
```

a phosphatidic acid

Table 6.4—*continued*
a. Complex lipids—*continued*

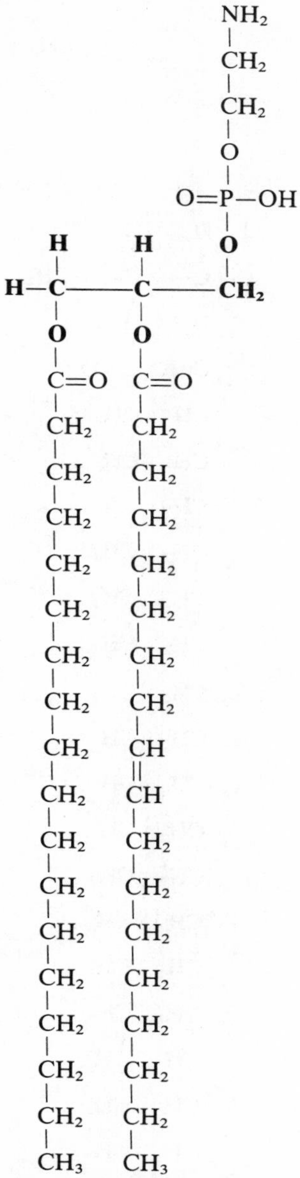

a phosphatidyl ethanolamine

Table 6.4—*continued*
a. Complex lipids—*continued*

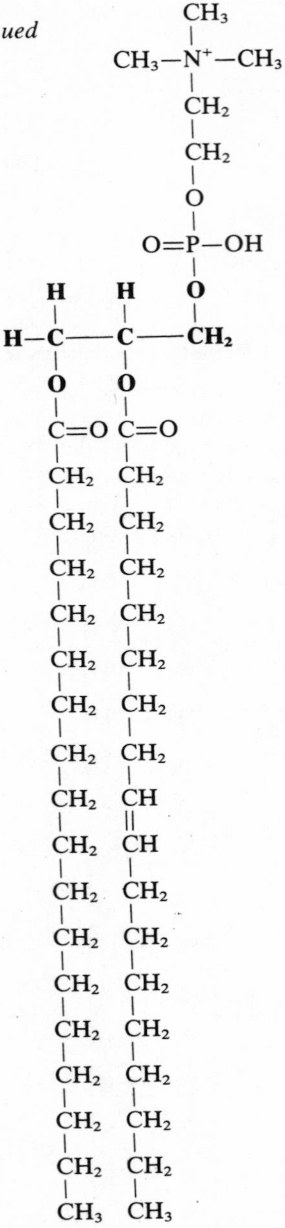

$$CH_3$$
$$|$$
$$CH_3-N^+-CH_3$$
$$|$$
$$CH_2$$
$$|$$
$$CH_2$$
$$|$$
$$O$$
$$|$$
$$O=P-OH$$
$$|$$
$$O$$

$$\begin{array}{ccc}
H & H & \\
| & | & \\
H-C & -C & -CH_2 \\
| & | & \\
O & O & \\
| & | & \\
C=O & C=O & \\
CH_2 & CH_2 & \\
CH_2 & CH_2 & \\
CH_2 & CH_2 & \\
CH_2 & CH_2 & \\
CH_2 & CH_2 & \\
CH_2 & CH_2 & \\
CH_2 & CH_2 & \\
CH_2 & CH & \\
CH_2 & CH & \\
CH_2 & CH_2 & \\
CH_2 & CH_2 & \\
CH_2 & CH_2 & \\
CH_2 & CH_2 & \\
CH_2 & CH_2 & \\
CH_2 & CH_2 & \\
CH_2 & CH_2 & \\
CH_3 & CH_3 & \\
\end{array}$$

a phosphatidyl choline

Table 6.4—*continued*
a. Complex lipids—*continued*

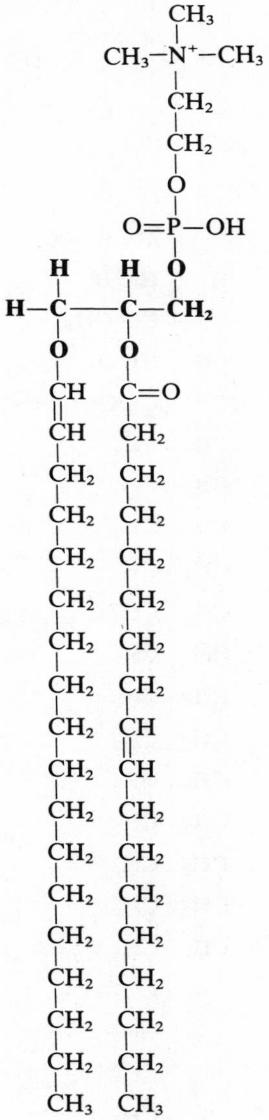

$$CH_3$$
$$CH_3-N^+-CH_3$$
$$CH_2$$
$$CH_2$$
$$O$$
$$O=P-OH$$

$$\begin{array}{cc} H & H \ O \\ H-C & C-CH_2 \\ O & O \end{array}$$

CH	C=O
‖	
CH	CH₂
CH₂	CH₂
CH₂	CH₂
CH₂	CH₂
CH₂	CH₂
CH₂	CH₂
CH₂	CH₂
CH₂	CH
CH₂	‖
CH₂	CH
CH₂	CH₂
CH₂	CH₂
CH₂	CH₂
CH₂	CH₂
CH₂	CH₂
CH₂	CH₂
CH₂	CH₂
CH₃	CH₃

a plasmalogen

Table 6.4—*continued*
a. Complex lipids—*continued*

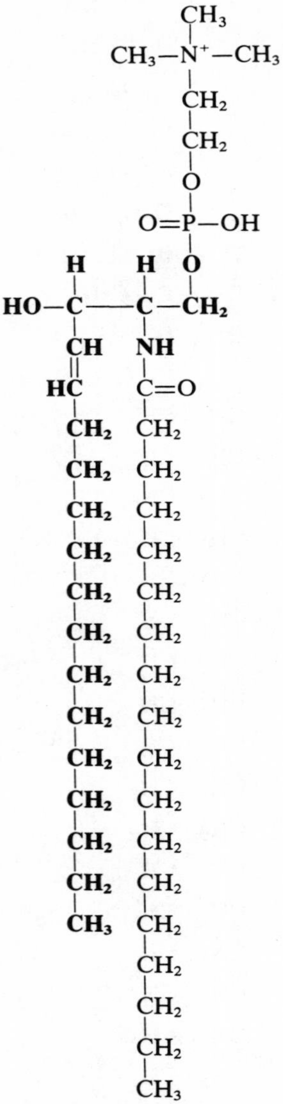

$$
\begin{array}{c}
CH_3 \\
| \\
CH_3-N^+-CH_3 \\
| \\
CH_2 \\
| \\
CH_2 \\
| \\
O \\
| \\
O=P-OH \\
\end{array}
$$

H	**H**	**O**
HO—C ——	**C**	**—CH₂**
CH	**NH**	
‖		
HC	**C**=O	
CH₂	CH₂	
CH₂	CH₂	
CH₂	CH₂	
CH₂	CH₂	
CH₂	CH₂	
CH₂	CH₂	
CH₂	CH₂	
CH₂	CH₂	
CH₂	CH₂	
CH₂	CH₂	
CH₂	CH₂	
CH₂	CH₂	
CH₃	CH₂	
	CH₂	
	CH₂	
	CH₂	
	CH₃	

a sphingolipid

Lipids

79

Table 6.4—*continued*
b. *Simple lipids*

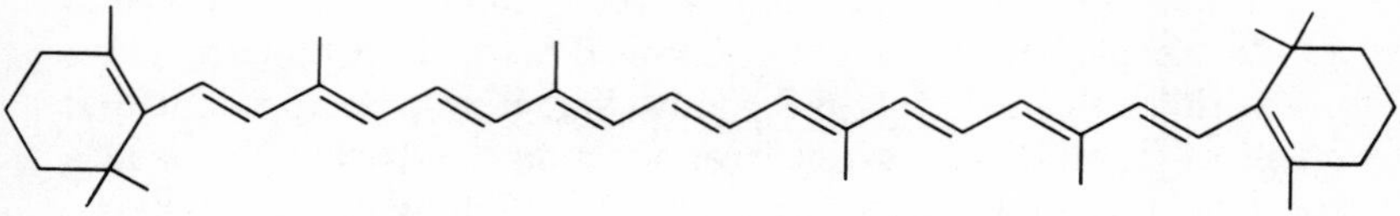

β-Carotene

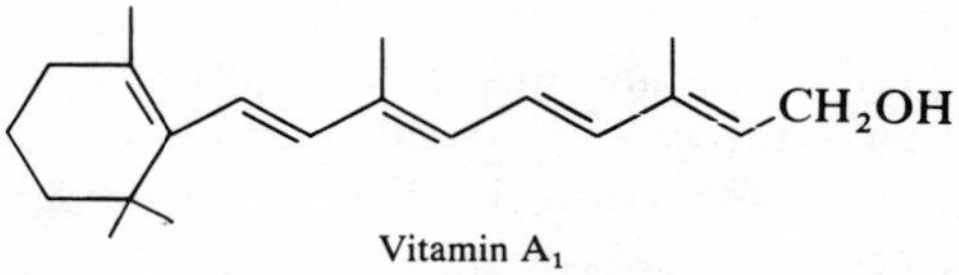

Vitamin A₁

two terpenes (derived from isoprene, $CH_2=C(CH_3)-CH=CH_2$)

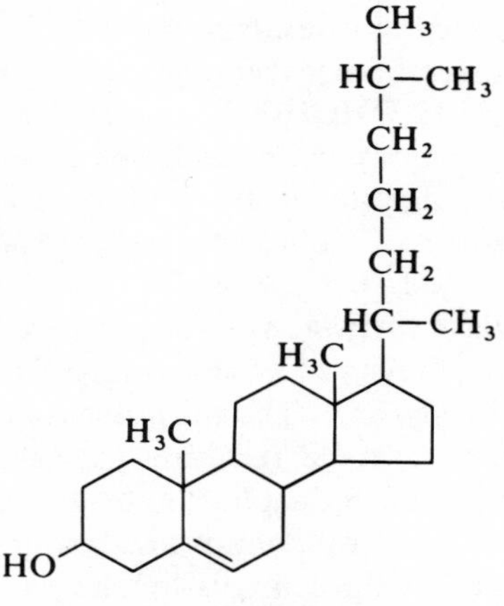

cholesterol, a steroid

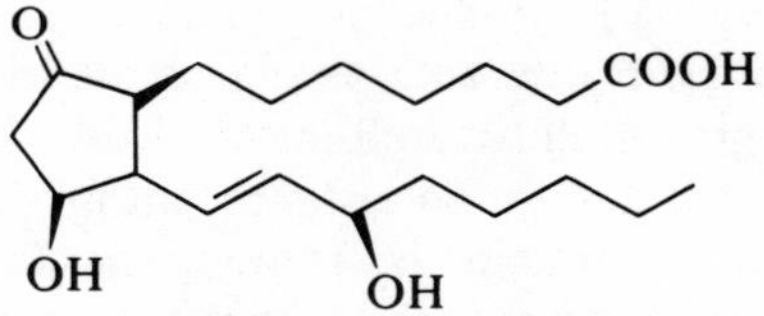

a prostaglandin (PGE₁)

By contrast with the building units of the proteins, nucleic acids and polysaccharides, the lipids include molecules that have nothing in common with each other except hydrophobicity: cholesterol and a fat molecule share few if any chemical features. (Compare Table 6.4 with Tables 3.1, 6.1, 6.3.) However, it will be apparent that much of Table 6.4 is given over to compounds that do have a common structural theme: they possess two long aliphatic chains, with the remainder of the molecule being drawn from a very wide range of possible structures.

In lipids of this latter type one at least of the long aliphatic chains tends to be a fatty acid, and this is often joined to the rest of the molecule by an ester linkage. Such linkages are susceptible to alkaline cleavage, otherwise known as saponification. Lipids susceptible to alkaline cleavage are often termed *saponifiable* lipids to distinguish them from those, such as cholesterol, that are not. The alternative terms are *complex* and *simple* respectively.

If one of the chains is attached via an ester linkage what of the other? Quite often this too is a fatty-acyl chain, also attached by esterification, as is indeed often true of the third, more variable, substituent. Three ester linkages require three –OH groups, and the compound glycerol ($CH_2OHCHOHCH_2OH$, propan-1,2,3-triol) is perfectly adapted to supply them, and thus to act as the spine on which the various components are attached. What might have seemed at first sight forbiddingly complicated structures can now be more readily understood. The *fats* are the simplest, since all three –OH groups of glycerol carry fatty-acyl chains. When there are only two fatty-acyl chains, they are found on the –OH groups of carbon atoms 1 and 2 of the glycerol. The third position can be occupied by a sugar (in which case the resulting molecule is known as a *glycolipid*) or more often by a phosphate group, to give rise to *phosphatidic acids*. These compounds are less important in themselves than as biosynthetic precursors for the *phosphoglycerides* in which the phosphate group itself has another group attached to it. The variety of possible groups that are found to be used for this purpose is very great (Table 6.5). The contents of the preceding chapters will have made us especially alert to the very wide range of types of non-covalent interactions that is made available through the diversity of groups illustrated in this Table.

Phosphoglycerides make up a significant proportion of the lipid content of animal cell membranes; among the commonest are those in which the substituent on the phosphate is one of the amino alcohols ethanolamine or choline (Table 6.5).

There are other ways in which the two aliphatic chains and the third, more variable component can be assembled. In the *plasmalogens*, for example, one of the chains is attached through an ether linkage rather than by esterification. More importantly, some lipids rely on compounds other than glycerol to provide the spine to which the aliphatic chains and the rest of the molecule are attached. In the *sphingolipids* one of the aliphatic chains is itself an integral part of the spine. (In the sphingolipid illustrated in Table 6.4 the spine is called *sphingosine*: the other possibilities are closely related to it and have recognizably similar names.) The second chain is a

Table 6.5. *Some of the common substituents on the phosphate group of phosphatidic acid. The* −OH *group in bold type is the one that joins the substituents to the phosphate group*

Phosphoglyceride	Alcohol component
Cardiolipin*	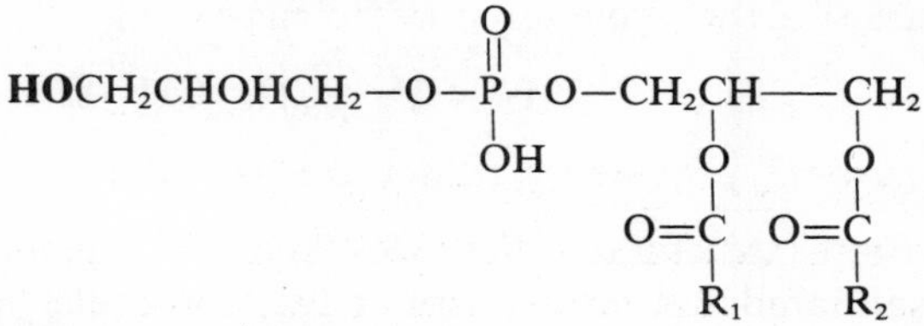 $\textbf{HO}CH_2CHOHCH_2-O-\overset{\displaystyle O}{\underset{\displaystyle OH}{P}}-O-CH_2CH-CH_2$, with $O=C-R_1$ and $O=C-R_2$
Phosphatidyl choline	$\textbf{HO}CH_2CH_2\overset{+}{N}(CH_3)_3$
Phosphatidyl ethanolamine	$\textbf{HO}CH_2CH_2NH_2$
Phosphatidyl glycerol	$\textbf{HO}CH_2CHOHCH_2OH$
Phosphatidyl inositol	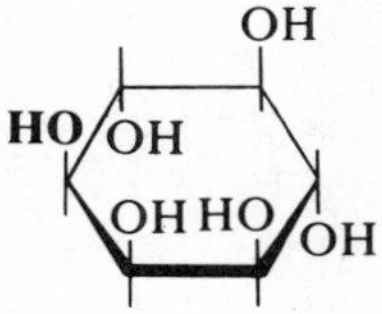 inositol ring with OH groups
Phosphatidyl serine	$\textbf{HO}CH_2CHNH_2COOH$
Phosphatidyl 3′-*O*-aminoacyl glycerol†	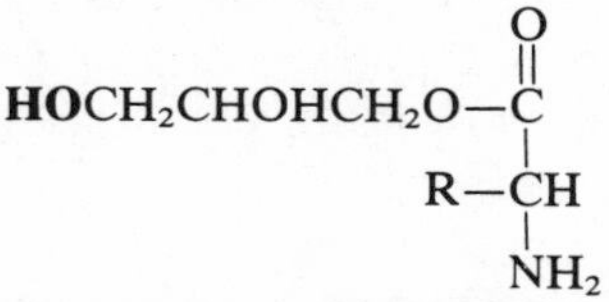 $\textbf{HO}CH_2CHOHCH_2O-\overset{\displaystyle O}{C}$, $R-CH$, NH_2

* R_1 and R_2 represent the alkyl chains of fatty acids.
† R represents the side chain of an amino acid.

fatty acid, as before, but it is attached by an amide link to the amino group possessed by compounds of the sphingosine type. (Amides are cleaved by alkali and thus the spirit of the definition of the term 'saponifiable' is preserved.) The third substitution is via an –OH group on the spine, also as before, and much the same range of types of substituents is observed as in the glycerol-based compounds.

The fatty-acid parts of all these molecules can be of many kinds. Chain lengths are commonly between 12 and 20 carbon atoms (Table 6.6). Even numbers are strongly preferred for reasons to do with the biosynthetic pathways of these compounds (Chapter 16). The chains are frequently unsaturated at one, two or three places, and since the unsaturation is almost always of the *cis* kind rather than *trans*, this has a dramatic effect on the shape of the molecule. Saturated alkyl chains of the general formula

$$CH_3-CH_2-CH_2-CH_2-CH_2-$$

are straight, while unsaturated chains, e.g.

$$CH_3-CH_2-CH=CH-CH_2-$$

are bent, provided that they are *cis* and not *trans* (Fig. 6.14).

These structural differences help to explain the fact that the saturated and *trans* forms of such molecules have higher melting points than *cis*-unsaturated forms: the bent chains do not pack together so well as do the straighter chains, and they are therefore more easily jerked apart by random thermal motion. As biologists, we are not very interested in melting as such, but we are very interested in the closely related phenomenon of *fluidity*. There is no

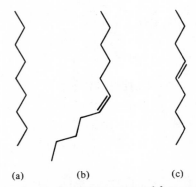

(a)   (b)   (c)

Fig. 6.14. *Cis–trans* isomerism in unsaturated fatty acids. (a) A saturated chain; (b) a *cis*-unsaturated chain; (c) a *trans*-unsaturated chain.

Table 6.6. *Fatty acids*

Carbon atoms	Common name	Systematic name	Structure
Saturated fatty acids			
12	lauric	n-dodecanoic	$CH_3(CH_2)_{10}COOH$
14	myristic	n-tetradecanoic	$CH_3(CH_2)_{12}COOH$
16	palmitic	n-hexadecanoic	$CH_3(CH_2)_{14}COOH$
18	stearic	n-octadecanoic	$CH_3(CH_2)_{16}COOH$
20	arachidic	n-eicosanoic	$CH_3(CH_2)_{18}COOH$
Unsaturated fatty acids			
16	palmitoleic		$CH_3(CH_2)_5CH=CH(CH_2)_7COOH$
18	oleic		$CH_3(CH_2)_7CH=CH(CH_2)_7COOH$
18	linoleic		$CH_3(CH_2)_4CH=CHCH_2CH=CH(CH_2)_7COOH$
18	linolenic		$CH_3CH_2CH=CHCH_2CH=CHCH_2CH=CH(CH_2)_7COOH$
18	*trans*-vaccenic		$CH_3(CH_2)_5CH=CH(CH_2)_9COOH$ *(trans)*
20	arachidonic		$CH_3(CH_2)_4CH=CHCH_2CH=CHCH_2CH=CHCH_2CH=CH(CH_2)_3COOH$

covalent bond to hold the lipids together, and so the ways in which the non-covalent properties of their hydrophilic parts can combine are profoundly influenced by the freedom of the individual molecules to move relative to one another. Really fluid lipid structures can constantly rearrange their constituents to provide new combinations to meet different circumstances, or to permit movement of some parts of the structure relative to others. The tighter the alkyl chains of the lipids can pack together, the firmer will be the structures that they form, while the looser the packing, the more fluid will these structures be. The degree of *cis* unsaturation in the lipids of an organism is under tight control. (Nature rarely bothers to make *trans*-unsaturated chains, perhaps because their shapes are so similar to the saturated ones.) Different parts of an organism may require quite different degrees of fluidity, and this can be achieved by differences in the compositions of the lipids. Alternatively, the degree of *cis* unsaturation can be altered to keep the degree of fluidity constant under different circumstances. The lipids near the hoof region of the leg of a reindeer are less saturated than those in more central parts of the body, reflecting the fact that, at the lower temperatures experienced by the extremities, the lipids must be prevented from congealing.

There are other saponifiable lipids which, though important in their specific roles, have less wide general significance. For example the *waxes* consist of esters formed between long-chain fatty acids and long-chain fatty alcohols: a wax is illustrated in Table 6.4.

We have now described the main types of option available to the cell in making up the complex, or saponifiable, lipid molecules. If we count up how many kinds of common fatty-acid chains there are, how many types of molecule can act as a spine, how many sorts of substitution there can be in the third position, and multiply these numbers together to give the total number of permutations, we find that there are about 10^4 possible structures. (Remember that these are only *monomers* that are used to make up the higher order, non-covalently associated, structures.) Note that the range of properties includes almost every type of non-covalent bond there is. Certain of the structures (notably phosphatidyl choline, phosphatidyl ethanolamine, cardiolipin, sphingomyelin and the non-saponifiable lipid cholesterol) predominate in many natural lipid structures, and we do not suggest that every one of the other possible permutations exists. However, a good number of them either are present in smaller amounts or could be called on for special requirements as these arise.

The non-saponifiable lipids are a diverse group of compounds (Table 6.4b). Many are long-chain molecules that are derived biosynthetically from the five-carbon unit isoprene and are called *terpenes*. The structures of others are dominated by the possession of four fused rings of carbon atoms, three of which are six-membered, and one of which is five-membered. Members of this group are known as the *steroids*. We shall see later that it is important not to forget that, although the particular way in which the structure is represented makes them look flat, leaflet-like compounds, in reality the rings have considerable thickness. There are many other types of non-saponifiable lipid, for example the *prostaglandins* and a range of fat-soluble vitamins, but the examples already given are probably sufficient to illustrate the diversity of structures available.

We will now consider some of the major uses of lipids in the body.

As mentioned when dealing with the carbohydrates, living organisms often lay down reserves of food. *Fats*, being totally insoluble, are excellent for this purpose, as they can be deposited as droplets in a convenient part of the cell. The more unsaturated the fatty-acyl chains, the more fluid the fat will be. Olive oil is a liquid at room temperature, unlike most animal fats, because of its unsaturation. The droplets will remain inert until the energy stored within them is required, when they can be degraded. The biosynthesis of fats is described in Chapter 16, their degradation in Chapter 12.

Biological *membranes* consist of, on average, 60% protein (by weight) and 40% lipid, a good deal of which is phosphorylated. Some of the protein molecules are enzymes, and in membranes from the appropriate parts of the cell they include many of the integrated multienzyme systems that catalyse the metabolic pathways that we shall describe in the next section of this book. Others of the protein molecules appear to have a role in promoting the transport of substances across the membranes (Chapter 21). Still others are concerned with the transport of reducing power and electrons in oxidative metabolism (see Chapter 8).

Most models of the gross anatomy of the membranes postulate a *bilayer* structure which arises as follows: the lipid fraction of the membrane material will be driven to associate by means of the hydrophobic interaction. All the monomers will try to align their non-polar chains for the maximum degree of side-to-side contact and will present their polar parts to the surrounding water. Contact between non-polar parts and water will be diminished and stability

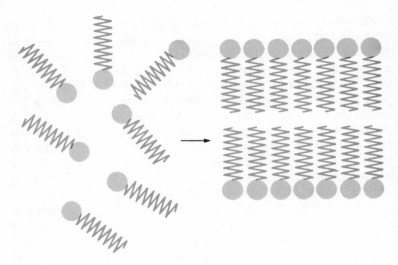

Fig. 6.15. The formation of lipid double layer.

will be enhanced if two of these structures form a sandwich (Fig. 6.15). Once again, therefore, we have a micelle (cf. Fig. 2.5), in this case a thin, flat one. As there is no covalent backbone the micelle can be very large. It can also be of quite complex structure (e.g. Fig. 6.16). It is particularly important to remember when considering membranes that any one component may be *mobile* and travel quite widely within the half of the bilayer structure in which it finds itself. Mobility will be controlled, as we have said already, by the degree of *cis* unsaturation of the aliphatic chains. The presence of cholesterol (which has a shape that does not fit tightly with aliphatic chains) has a similar effect on mobility as does *cis* unsaturation. Fig. 6.16 indicates the sort of ways in which the different hydrophobic parts of the lipid could pack together.

If the association between the lipid monomers is largely of the one, hydrophobic variety, what is the role of the various types of other non-covalent property possessed by the hydrophilic parts of the lipid? Some will be needed to stabilize the interface between the micelle and water; some will interact specifically with each other to give rise to regions with particular geometrical or chemical properties, and some will, singly or in combination, form binding sites for the large numbers of different proteins that give the membranes their unique properties.

Correspondingly, the proteins are likely to have stretches of hydrophobic amino acids which will penetrate the non-polar part of

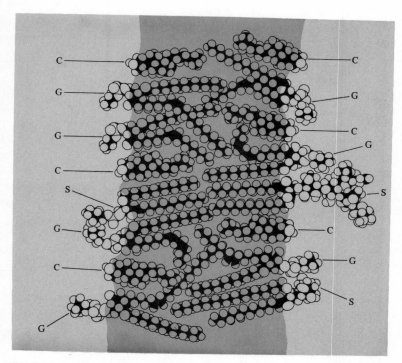

Fig. 6.16. An idealized view of the packing of lipids in a membrane bilayer. The inner, hydrophobic region is formed by the aggregation of the hydrocarbon chains of the complex lipids and cholesterol. The bends shown in the hydrocarbon chains are all due to *cis* unsaturation (p. 82): in reality the saturated parts of the chains, which are shown here as straight, would show some additional bends as a result of random thermal motion. Such bending would help to reduce the size and number of the gaps shown between the different molecules. Where it occurs, the additional bending is believed to be most significant near the centre of the membrane. The complex lipids are of both the glycerol-bridged and sphingosine-derived types (Table 6.4). The overall appearance of the two types (which are indicated by G and S respectively) is very similar, while both are noticeably unlike cholesterol (C). The hydrophilic regions consist of a variety of polar substituents of the complex lipids, such as phosphoryl choline, phosphoryl ethanolamine and various carbohydrate structures, together with the –OH group of cholesterol. The particular region shown does not include any protein molecules.

the layer. It will then be possible for the polar parts of the protein to add their own hydrophilic contribution to stabilizing the interface with the surrounding water (Fig. 6.17). (Non-covalently bound lipoprotein complexes of this type are also known in other situations

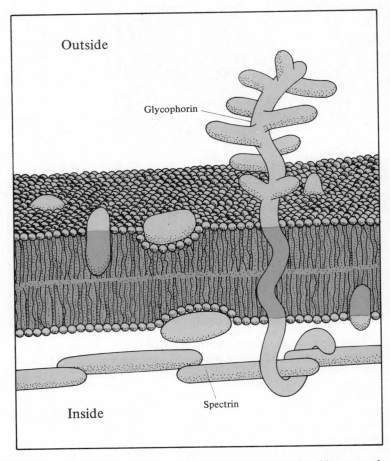

Fig. 6.17. An idealized view of a cell membrane (in this case, the membrane of a red blood cell). The lipid bilayer is visible, together with a number of proteins. Some are quite hydrophilic and are merely adsorbed onto the surface of the bilayer. Others have strongly hydrophobic parts, which are found deep in the lipid layer. The glycoprotein known as glycophorin is one of the characteristic molecules of the cell's surface. The N-terminal portion of its sequence (the upper part in the Figure) has a number of carbohydrate chains covalently attached, which are shown as branches to the main chain of the protein. This hydrophilic N-terminal region protrudes into the extra-cellular medium. The portion of the sequence that traverses the lipid layer has a great many hydrophobic residues. The C-terminal portion of glycophorin interacts, it is conjectured, with spectrin, a structural protein of the cell's interior.

than the membrane, for example in the transport of fats between different tissues of the body.)

Although little is known as yet about the fine structure of the membrane surface there is reason to suppose that it will be a *mosaic* of protein assemblies embedded in the bilayer, some retained in fixed positions, others free to some extent to wander along the surface of the membrane.

Recent studies with the electron microscope have led to some controversy over the detailed morphology of membranes and over the universality of the bilayer structure. However, the picture in Fig. 6.17 is probably quite close to the truth in many important instances and sufficient for a discussion of the principal function of the membranes – which is to allow, or, when necessary, to prevent, the passage of substances from one side of them to another. We shall return to this subject in Chapter 21, by which time we will have enumerated some of the principal types of compartment within the cell, and have discussed the more important classes of substances that have to be transported either between compartments, or between the inside and the outside of the cell.

Section II
Metabolism

7 Energy and biochemical reactions

It is everyday experience that all processes can be divided into two categories, those that tend to occur by themselves and those that must be driven by the expenditure of energy. Water running downhill is an example of the first class, a vehicle moving uphill is an example of the second. Furthermore we know that it is processes of the first class that yield the energy to drive those of the second (a water wheel may be used to pull a trolley uphill). This division applies just as much at the molecular level, and it is of great importance in the design of living systems. In particular, by analogy with the mechanical examples cited *it is possible to couple the chemical energy yielded by a process of the first type to drive a chemical or physical process of the second type.*

Living matter depends on this possibility. Organisms exploit it to bring about syntheses of compounds and structures which could not otherwise be made and also in order to couple chemical energy with mechanical work. (Examples of coupling between chemical and mechanical processes are muscular contraction and the forced transport of solutes into areas of higher concentration.) Just as the exploitation of energy coupling is a characteristic of life at the molecular level, so the consequences of that coupling – the ability to grow, to move and to organize – are the characteristics by which we recognize life at the macroscopic level.

To assess the place of any particular process in the economy of the living state we need to know in which category it lies and to have some idea of the amount of energy that could be obtained from it or that would be needed for it. There exists a useful thermodynamic quantity to help us. Consider a process $A \rightleftharpoons B$ taking place at constant temperature and pressure – the conditions under which living matter normally operates. The maximum useful energy produced or consumed is known as ΔG, the change in the so-called Gibbs free energy of the reacting system. By convention, if ΔG has a negative value when going from A to B, energy may be obtained from the process and the process $A \rightarrow B$ is of the first, energy-yielding, category. If the value is positive then $A \rightarrow B$ is of the second, energy-requiring, category.

Clearly if ΔG is negative for A→B then, as the process is reversible, it will be positive for B→A. This convention is not so illogical as it may seem: if we obtain energy (negative ΔG) from the process the system itself must lose it.

Thus the sign of ΔG tells us in which category the process lies. Our second requirement, to know how much energy is involved, is met by its magnitude. The greater the negative magnitude of ΔG, the more energy is to be had. The greater the positive magnitude, the more is needed.

It must not be thought that whenever ΔG is negative the process will always and instantaneously occur. To return to the water analogy, consider a reservoir a certain height above the surrounding countryside. There is clearly a strong tendency ('high negative ΔG') for the water to run down. Whether it actually does so or not, and the rate if it does, depend on factors unrelated to this value. If the dam is effective the water will never run down; if there is an infinitesimal hole it may do so but at an imperceptible rate; if the dam is destroyed it will do so with great violence.

Thus returning to chemical reactions we can speak of a reaction being energetically favoured (negative ΔG) but kinetically hindered (see p. 47). The violent combustion of our bodies in air has a large negative ΔG. We survive because at room temperature there is kinetic hindrance of the process – the so-called 'potential barrier' (see Fig. 5.2).

The value of ΔG clearly depends on the quantities of the components present. A cupful of water descending from the reservoir could not generate the same amount of energy as a million gallons.

This dependence on quantity leads to a very valuable relationship between ΔG and other parameters of chemical reactions. Consider the reversible reaction A $\rightleftharpoons$ B in which A → B is energetically favoured over B → A. If we start with equal concentrations of A and B the overall reaction will proceed from left to right and the concentration of B will build up while that of A will decrease. The effective ΔG of B → A will increase in negative value as the concentration of B increases, and the effective ΔG of A → B will fall in negative value as the concentration of A falls. Eventually the concentrations will reach a point at which the ΔG's for the forward and backward reactions are equal. At this point no further change in the overall concentrations of A and B will occur.

What we have described is, of course, the approach to equilibrium of the reversible reaction A $\rightleftharpoons$ B. It should, therefore, not be surprising to find a relationship between ΔG and the equilibrium constant K, since the latter quantity simply expresses the ratio between the products of the forward and of the backward reactions.

The relationship is

$$\Delta G^0 = RT \ln K, \tag{1}$$

where ΔG^0 is the so-called *standard change* in the Gibbs free energy of the forward reaction at 25 °C. The term 'standard change' implies that the effective concentration of all reactants is taken as molar (gases at one atmosphere). R is called the gas constant, T is the absolute temperature.

If ΔG is expressed in joules per mole, this expression reduces to

$$\Delta G^0 = -5730 \log_{10} K. \tag{2}$$

This equation considerably extends the uses to which a knowledge of the value of ΔG^0 can be put, since it allows us to calculate the equilibrium constant and hence the extent to which a reaction will proceed, as well as the overall direction and the possible yield of chemical energy. Alternatively, if the equilibrium constant is already determined, we may at once calculate ΔG^0.

Values of ΔG^0 are not known for all biochemical reactions. Even where values are known many have been determined in free solution and not under conditions realistically reproducing those in the cell. Nonetheless the values that are available probably do not misrepresent the position too seriously. They have proved so useful that we may expect further and more accurate values of ΔG^0 to become available as time goes on.

Energy in metabolic reactions

To illustrate the use of the ideas contained in the previous pages, let us consider some biochemical reactions drawn from later chapters. First a reaction in which $\Delta G^{0\prime}$ is fairly near zero.*

$$\text{Glucose-6-phosphate} \rightleftharpoons \text{fructose-6-phosphate} \tag{3}$$
$$\Delta G^{0\prime} = +2.1 \text{ kJ/mol}$$

is a reaction of some importance in both the synthesis and degradation of carbohydrates (pp. 164 and 126). From equation (2)

$$\log_{10} K = -\frac{2.1}{5.73} = -0.367 = \bar{1}.633,$$
$$K = \text{antilog } \bar{1}.633 = 0.43.$$

---

* $\Delta G^{0\prime}$ is ΔG^0 corrected to pH 7 rather than pH 0 (the pH at which the molarity of $H^+ = 1$) as would be demanded by the definition of ΔG^0 for those reactions involving H^+. The choice of pH 7 is, of course, much more realistic.

Thus at equilibrium at pH 7 there will be 0.43 times as much fructose-6-phosphate as glucose-6-phosphate. If there is 1 μmol of glucose-6-phosphate present there will be 0.43 μmol of fructose-6-phosphate at equilibrium.

There are important consequences of the fact that the equilibrium constant is near to one. Biochemical reactions usually exist to bring about the net formation of a compound which may be required either for itself or as the starting material of a further process (Chapter 9). Reactions with equilibrium constants near unity can, in distinction to some that we shall discuss below, bring about a useful net formation of product in either direction. If one were to add 14.3 μmol of fructose-6-phosphate to the system, a net formation of 10 μmol of glucose-6-phosphate would result. If one were to add 14.3 μmol of glucose-6-phosphate there would be a net synthesis of 4.3 μmol of fructose-6-phosphate, a reasonable amount, even though the reaction in this direction is slightly disfavoured energetically. Thus the reaction may readily be used in degradation of carbohydrate, which is found (p. 126) to require the reaction from left to right, and in synthesis, which is found (p. 164) to require the reaction from right to left.

Let us turn to another reaction in carbohydrate metabolism, the equilibrium between dihydroxyacetone phosphate and glyceraldehyde-3-phosphate.

$$
\begin{array}{ccc}
CH_2OH & & CHO \\
| & & | \\
CO & \rightleftharpoons & CHOH \\
| & & | \\
CH_2O\circledP & & CH_2O\circledP
\end{array}
\qquad (4)
$$

$$\Delta G^{0\prime} = +7.7 \ kJ/mol.$$

The equilibrium constant is

$$\text{antilog} -\frac{7.7}{5.73} = \text{antilog } \bar{2}.66, \text{ i.e. about } \frac{1}{22}.$$

At equilibrium there will be 22 μmol of dihydroxyacetone phosphate to one of glyceraldehyde-3-phosphate. Now we shall see (p. 127) that carbohydrate breakdown requires the reaction to proceed from left to right. Although more extreme examples are to come, the 'wrong' product seems even here to be very much favoured by the equilibrium constant. How can the process be used for a significant net formation of glyceraldehyde-3-phosphate and thus a significant breakdown of carbohydrate? The answer is that an enzyme system exists which is ready to take such glyceraldehyde-3-phosphate as there is and convert it to the next product in the chain of reactions leading to carbohydrate breakdown. In an attempt to restore equilibrium more glyceraldehyde-3-phosphate will be formed to replace it. So long as the enzyme system continues to tap off the glyceraldehyde-3-phosphate as it is produced, a useful net formation of product will occur. Thus virtually any quantity of dihydroxyacetone phosphate can be transformed to glyceraldehyde-3-phosphate.

It is necessary, of course, that the overall equilibrium constant of the process which removes glyceraldehyde-3-phosphate should favour removal rather than synthesis. Putting this in free energy terms, the $\Delta G^{0\prime}$ of the removal process must be sufficiently negative overall to overcome the positive $\Delta G^{0\prime}$ of equation (4).

Another most important example of the use of this 'tapping off' process to control the direction of net formation of product is given on p. 141.

Now for a reaction with a still higher positive value of $\Delta G^{0\prime}$:

Ribulose-5-phosphate + phosphate $\rightleftharpoons$ ribulose-1,5-bisphosphate

(5)

$$\Delta G^{0\prime} = +10.1 \text{ kJ/mol},$$

$$K = \text{antilog} -\frac{10.1}{5.73} = 1.73 \times 10^{-2}.$$

Ribulose-1,5-bisphosphate is to be used for the synthesis of compounds such as carbohydrates. It is therefore not practical to attempt, by analogy with the removal of glyceraldehyde-3-phosphate in equation (4), to achieve a net production of bisphosphate by tapping it off to a lower energy state than the start of the process. This is because the available free energy of the carbohydrate that is to be formed from the bisphosphate is higher than that of ribulose-5-phosphate, so that the *overall* $\Delta G^{0\prime}$ would not be negative.

Furthermore, the equilibrium constant makes it impractical to achieve a useful net formation of the bisphosphate by increasing the concentration of the reactants on the left of the equation, as can be done in the case of equation (4). How then is this reaction and the many others like it, equally vital to living matter and equally unfavourable on energetic grounds, to be brought about? The answer is, as we saw on p. 93, that processes may be coupled. In other words, the problem is solved by *coupling* the processes of high positive ΔG to those of high negative ΔG.

Before we can consider the systems used for coupling we must round off this survey of the energetics of biochemical reactions by quoting an example of a reaction with a high negative ΔG, the equilibrium between phospho-*enol*pyruvate and pyruvate (see p. 129).

$$
\begin{array}{ccccccc}
\text{CH}_2 & & & & \text{CH}_3 & & \\
\| & & & & | & & \\
\text{CO}\textcircled{P} & + \text{H}_2\text{O} & \rightleftharpoons & & \text{C}=\text{O} & + \text{P}_i & \qquad (6) \\
| & & & & | & & \\
\text{COOH} & & & & \text{COOH} & &
\end{array}
$$

$$\Delta G^{0\prime} = -55.4 \text{ kJ/mol.}$$

This is a step in the degradation of carbohydrate; it is clear that there is at least the possibility of a useful energy yield in this process which might be coupled to reactions of high positive ΔG. This possibility, as we shall see later, is made a reality. For the moment we note that the equilibrium constant is antilog $\frac{55.4}{5.73} = 4.6 \times 10^9$ in favour of the forward (left to right) reaction.

We can use this example to explain the misuse so often made in biochemistry of the terms 'reversible' and 'irreversible' as applied to metabolic reactions. These terms are not to be taken to refer to the reversibility or otherwise of the *mechanism* of the reaction. They refer only to the ease or otherwise of achieving net formation of product in both directions. The reaction fructose-6-phosphate $\rightleftharpoons$ glucose-6-phosphate was an example of a 'reversible' reaction in this sense. The reaction phospho-*enol*pyruvate $\rightleftharpoons$ pyruvate (note the double arrows) is an 'irreversible' reaction in the sense that one would not seek (without some external energy source) to use the process for the net synthesis of phospho-*enol*pyruvate.

Some people find attempts to apply thermodynamics to metabolic phenomena rather bogus. They point to the fact that we often find, as in equation (4), that the known direction in which the process can be operated in the cell is opposite to that indicated by a simple calculation based on ΔG. We advanced on p. 97 the argument that such processes are pulled over against the natural thermodynamic tendency by the rapid conversion of the products to other substances. Critics of the use of ΔG find this unsatisfactory, since it appears to allow us to accept the verdict of ΔG whenever we wish, and talk our way around it when we do not. Nonetheless, we have felt it appropriate to make references to thermodynamic quantities in this book. They provide the only framework in which it is possible to discuss the kind of issues dealt with in this chapter. The objection just cited seems to us to lack force because one could equally well claim the case of reaction (4) as an instance in which thermodynamics has drawn attention to, and helped to quantify, a biologically very important phenomenon: the use of further conversion of products in a reaction of negative ΔG to influence the net direction of reaction.

Energy sources and energy coupling

A chemical reaction is in a sense solely concerned with electrons. If the electrons that maintain the structure of chemical compounds shift permanently from one configuration to another, we say that a chemical reaction has taken place. If the potential energy of the first configuration is higher than that of the second (more negative ΔG) then the reaction could be made to yield energy. If the reverse is true it will require energy. If we imagine ourselves designing a living organism we must search for a source of energy to couple to its essential energy-requiring processes. We are thus really looking for electrons in a state of high potential energy (high negative ΔG) and a sink of lower potential energy (less negative ΔG) into which we can put them.

Save for a few trivial exceptions there is only one *primary source* of electrons at a high potential, able to fall to a lower, which is suitable for coupling to the energy need of living organisms.

These high-potential electrons occur in the photosynthetic process (Chapter 14), carried out in a 'solid-state' device found in plant cells which is analogous to but much more sophisticated than a transistor. The device, which is known as the chloroplast, consists of an assembly of proteins, prosthetic groups and lipids. A quantum of light is absorbed, and its energy of red light is given up in promoting an electron to a higher potential energy state. The electron then falls back to its original state. In a way which is not fully understood, the energy of the fall is coupled to the synthesis of certain so-called 'high-energy' compounds.* These compounds, which are discussed below, can undergo energy-yielding reactions, which can be coupled to energy-requiring reactions, and they can be regarded as a *secondary source* of free energy. By the use of this small group of 'high-energy' compounds, a large number of molecules can be made whose synthesis would otherwise be energetically disfavoured to a greater or lesser extent; for example carbohydrate and fat (Chapters 15 and 16). These products then form a *tertiary source* of

---

* 'High energy' is a despised term in some quarters and does indeed have its drawbacks. However, it is a useful shorthand term and as long as it is used with care it is worthy to be retained.

electrons at a useful potential energy level. The plants themselves use such photosynthetically produced compounds as food reserves (p. 66). Animals obtain these compounds either by consuming vegetable matter directly or, if they are carnivorous, via the food chain.

Given that we have, in the food reserves thus formed, a source of electrons at a high configurational energy, we must now look for a sink at a lower energy state – that is to say an electron acceptor – into which they may fall, so that the energy of the fall may be coupled to endergonic processes. Simple processes like

$$\text{Fe}^{2+} \rightleftharpoons \text{Fe}^{3+} + \text{e}^- \qquad (7)$$

remind us that to donate an electron is to be oxidized. Although other molecules exist which are capable of acting as acceptors, molecular oxygen owes its almost unrivalled ability to act as an electron acceptor to certain peculiarities of the way in which its electron shells are filled.

Oxidative metabolism is, therefore, a very important source of energy. Oxidation reactions of the type

$$\text{AH}_2 + \tfrac{1}{2}\text{O}_2 \rightleftharpoons \text{A} + \text{H}_2\text{O} \qquad (8)$$

can easily yield a $\Delta G^{0\prime}$ of more than 100 kJ/mol.† This would be more than enough to drive the endergonic processes of the cell, equation (5) for example. Cells that utilize oxidative reactions as an energy source also have a 'solid-state' device, this time called the mitochondrion (Chapter 21), to trap this energy by synthesizing high-energy compounds, in a process called oxidative phosphorylation. The components of the mitochondrion are, as far as the different functions permit, similar to those of the chloroplast. The high-energy compounds produced are either identical or closely related.

It is now time to describe these high-energy compounds and the way in which they are used in energy coupling.

ATP and energy coupling

Both the photosynthetic and the oxidative processes produce the molecule adenosine triphosphate (ATP). The structure is given in

---

† You may be accustomed to treating oxidation reactions differently from all others and expressing the energy levels in terms of standard electrode potentials. These potentials are directly convertible to ΔG^0 (a change in standard potential of $+1$ V for 1 g-equiv. of electrons at pH 7 corresponds to a $\Delta G^{0\prime}$ of 96.6 kJ/mol). In order to stress the unity of all energy-yielding reactions the one scale, ΔG^0, will be used throughout this book.

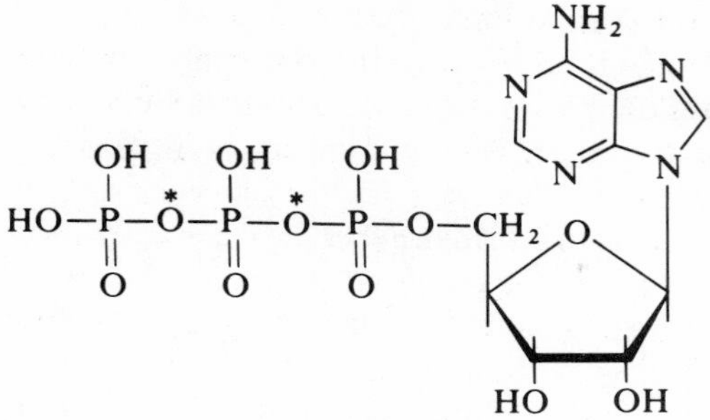

Fig. 7.1. The chemical structure of ATP.

Fig. 7.1. It is a property of the phosphoric-anhydride bond (marked with an asterisk in the figure) that it has a considerable free energy of hydrolysis.

$$R{-}O{-}\overset{\overset{\displaystyle O}{\|}}{\underset{\underset{\displaystyle O^-}{|}}{P}}{-}O{-}\overset{\overset{\displaystyle O}{\|}}{\underset{\underset{\displaystyle O^-}{|}}{P}}{-}O^- + H_2O \rightleftharpoons R{-}O{-}\overset{\overset{\displaystyle O}{\|}}{\underset{\underset{\displaystyle O^-}{|}}{P}}{-}O^- + \text{phosphate}$$

$$\Delta G^{0\prime} = \text{about } -30 \text{ kJ/mol.}$$

The loss of the terminal phosphate group yields adenosine diphosphate (ADP), while the loss of two phosphate groups gives adenosine monophosphate (AMP).

It is difficult to arrive at values for the $\Delta G^{0\prime}$ in the cell of the two reactions

$$ATP + H_2O \rightleftharpoons ADP + P_i \qquad (9)$$

and

$$ATP + H_2O \rightleftharpoons AMP + PP_i \qquad (10)$$

that are completely satisfactory. These reactions are affected by a number of factors, the precise effect of which is difficult to determine. To take a single example, all the reactants and products in equations (9) and (10) have a strong tendency to form complexes with Mg^{2+}, an ion which occurs in appreciable concentrations in physiological fluids. Since the free energy of hydrolysis of the Mg^{2+} complexes will be quite different from the free energy for the parent compounds, some correction ought to be made to take this factor into account. Unfortunately, this correction cannot be easily made with any precision. Because of this and other, similar difficulties, the values that we use in this book (−31.1 kJ/mol for equation (9) and

-31.9 kJ/mol for equation (10)) are to be regarded as approximations only. They are unlikely to be so wrong as to invalidate any of the qualitative conclusions that we base on them.

During the photosynthetic or oxidative synthesis of ATP (*photosynthetic phosphorylation* or *oxidative phosphorylation*) the energy of the falling electron is somehow used to drive reactions such as the one shown in equation (9) from right to left. The resulting ATP can now move to a site in the cell at which, for example, reaction (5) is to proceed.

We can, therefore, combine equations (5) and (9) to obtain:

Ribulose-5-phosphate + ATP $\rightleftharpoons$

$$\text{ribulose-1,5-bisphosphate} + \text{ADP} \qquad (11)$$
$$\Delta G^{0\prime} = 10.1 + (-31.1) = -21 \text{ kJ/mol}.$$

The equilibrium constant is now antilog $\frac{21}{5.73} = 4.6 \times 10^3$ in favour of the *formation* of ribulose-1,5-bisphosphate. In contrast to the situation when equation (5) is taken alone, the synthesis of the bisphosphate now proceeds almost to completion.

Coupling of this sort, with high-energy compounds as intermediate carriers of the energy, takes place in very many chemical reactions. The product of the energy-requiring reaction is not always a phosphorylated compound even when the reaction is being driven by a high-energy phosphate (p. 175). Coupling with the breakdown of high-energy compounds also drives mechanical processes that are endergonic, as well as chemical ones (p. 34).

ATP is not the only 'high-energy' compound and Table 7.1 lists a number of others. ATP is, however, the principal medium of energy exchange.

Phospho-*enol*pyruvate (equation (6)) can now be seen as an example of a high-energy compound. It is not generally used as such directly, but is used indirectly, through the manufacture of ATP.

$$\text{PEP} + \text{ADP} \rightleftharpoons \text{pyruvate} + \text{ATP}. \qquad (12)$$

The overall energy change is obtained by combining equation (6) together with equation (9) written in the reverse direction. (The reverse of equation (9) necessitates a change of sign in its $\Delta G^{0\prime}$.)

$$\Delta G^{0\prime} = -55.4 + 31.1 = -24.3 \text{ kJ/mol}.$$

Thus there is a net formation of ATP.

This example also serves to point out that there are other means of producing ATP than photosynthetic or oxidative phosphorylation. This third type of process, in which the phosphorylation of ADP arises as part of the mechanism of a metabolic

Table 7.1. *High-energy compounds*

Type	Common examples	See page
Pyrophosphates 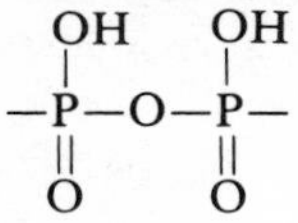	ATP, ADP; other nucleoside di- and triphosphates	34, 112, 125, etc.

$$\begin{array}{cc} OH & OH \\ | & | \\ -P-O-P- \\ || & || \\ O & O \end{array}$$

Type	Common examples	See page
Acyl phosphates 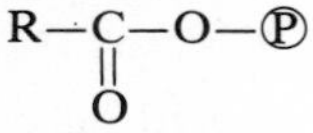	Glyceric acid-1,3-bisphosphate	128

$$R-\underset{\underset{O}{||}}{C}-O-\textcircled{P}$$

Type	Common examples	See page
Enol phosphates 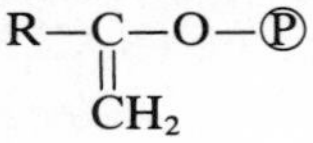	Phospho-*enol*pyruvic acid	127

$$R-\underset{\underset{CH_2}{||}}{C}-O-\textcircled{P}$$

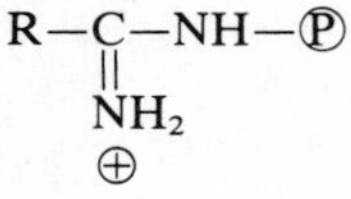

Type	Common examples	See page
Thioesters 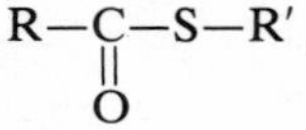	Acetyl CoA	134

$$R-\underset{\underset{O}{||}}{C}-S-R'$$

Guanidine phosphates   Creatine phosphate

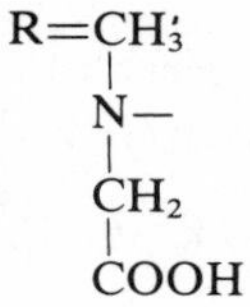

$$R-\underset{\underset{NH_2}{||}}{C}-NH-\textcircled{P}$$
$\oplus$

$R=CH_3^+$
$\quad | $
$\quad N- $
$\quad | $
$\quad CH_2 $
$\quad | $
$\quad COOH $

used as energy store in muscle

reaction, is called *substrate-level phosphorylation*. Although substrate-level phosphorylations are not of such general significance as the other two types, important examples do exist (pp. 128 and 136).

You may now realize that we had some time ago an example of what is in a sense an energy coupling. The tapping-off of glyceraldehyde-3-phosphate to a lower energy state employs a process of negative $\Delta G^{0\prime}$ to drive reaction (4) which has a positive $\Delta G^{0\prime}$. The process of negative $\Delta G^{0\prime}$ is the series of reactions (Chapters 10 and 11) which accomplish the stepwise oxidation of the triose to three molecules of carbon dioxide.

Coupling of oxidative and reductive reactions

It is possible to couple oxidative and reductive reactions without the intervention of ATP, or of any of the other high-energy compounds that we have so far considered.

For example, in fat metabolism (Chapters 12 and 16) certain derivatives of β-keto and β-hydroxy acids must be interconverted. This reaction can be written formally as:

$$CH_3(CH_2)_n COCH_2COOR + 2H \rightleftharpoons$$

$$CH_3(CH_2)_n CHOHCH_2COOR \quad (13)$$

$$\Delta G^{0\prime} = -42 \text{ kJ/mol.}$$

The nature of, and need for, the R groups are explained fully in Chapters 12 and 16.

The hydrogen is donated from the specialized, hydrogen-carrying molecule nicotinamide adenine dinucleotide phosphate (NADP). (The structure of this molecule is shown in Table 3.2, p. 26.)

The donation of hydrogen may be written as

$$NADPH_2 \rightleftharpoons NADP + 2H \quad (14)$$

$$\begin{array}{cc} \text{(reduced} & \text{(oxidized} \\ \text{form)} & \text{form)} \end{array}$$

$$\Delta G^{0\prime} = +17.2 \text{ kJ/mol}$$

and so the overall, coupled process is

$$CH_3(CH_2)_n COCH_2COOR + NADPH_2 \rightleftharpoons$$

$$CH_3(CH_2)_n CHOHCH_2COOR + NADP \quad (15)$$

$\Delta G^{0\prime}$ for equation (15) is about -25 kJ/mol and the reaction readily proceeds from left to right.

When reaction (13) is operated from right to left, a hydrogen *acceptor* is required. NADP could in theory be used for this purpose, but in fact the closely similar molecule nicotinamide adenine dinucleotide (NAD), which differs only very slightly in structure from NADP (see Table 3.2), is normally used instead.

The structures of NAD and NADP are, in fact, so similar that the energetics of reactions that involve them would not be significantly different if one replaced the other. We can therefore rewrite equation (15) to represent the use of NAD as an hydrogen acceptor:

$$CH_3(CH_2)_n CHOHCH_2COOR + NAD \rightleftharpoons$$

$$CH_3(CH_2)_n COCH_2COOR + NADH_2 \quad (15a)$$

$$\Delta G^{0\prime} = +25 \text{ kJ/mol.}$$

How is reaction (15a) driven from left to right in the face of this unfavourable $\Delta G^{0\prime}$? As is usual when an energetically unfavourable reaction has to be operated, it is *coupled* to one that is energetically favourable. The reaction that is used for coupling in the present case is

$$NADH_2 + \tfrac{1}{2}O_2 \rightleftharpoons NAD + H_2O \qquad (16)$$
$$\Delta G^{0\prime} = -218 \text{ kJ/mol.}$$

(Reaction (16), which is a multistep process, is discussed at length in Chapter 8.)

If we combine equations (15a) and (16) we obtain an example of a general type of oxidative reaction found in tissues.

$$CH_3(CH_2)_n CHOHCH_2COOR + \tfrac{1}{2}O_2 \rightleftharpoons$$
$$CH_3(CH_2)_n COCH_2COOR + H_2O \qquad (17)$$
$$\Delta G^{0\prime} = -193 \text{ kJ/mol.}$$

There is now no difficulty at all in operating the process from left to right. Indeed, there is free energy to spare and some of it is used to bring about the phosphorylation of three molecules of ADP to ATP. We have mentioned this *oxidative phosphorylation* of ADP before, and will return to it in Chapter 8.

In contrast to $NADH_2$, $NADPH_2$ does not generally enter oxidative phosphorylation, but is almost always used solely as a hydrogen donor, as in equation (15). As we shall see in Chapter 8, it is produced during the oxidation of only a few substrates and provides a convenient means of interrupting the oxidation of AH_2 at an early stage so that the reducing power can be used elsewhere rather than for reducing molecular oxygen as in equation (16). In fact it turns out to be a general rule that the reducing power of NADP(reduced), in distinction to that of NAD(reduced), is coupled to synthetic reactions (pp. 155, 167 and 172). We may note incidentally that reduced NADP is the other immediate product, beside ATP, of the fall of the electron in photosynthesis. This fact further emphasizes the role of reduced NADP in synthesis, since reduced NADP and ATP are used directly in the synthesis of carbohydrate in the plant (see Chapter 15).

We can summarize the preceding few pages by saying:

(1) The change in the Gibbs free energy, ΔG, is a valuable measure of the availability of, or need for, energy in a reaction.

(2) ΔG is related to the equilibrium constant and this makes it possible to use the energy term to calculate the direction in which the net reaction tends to go.

(3) Living matter operates energetically disfavoured reactions by coupling them with energetically favoured ones. The fall of a light-energized electron back to its ground state and the fall of electrons to molecular oxygen are the principal among the fundamental energy-yielding processes. These processes produce 'high-energy' compounds. The high-energy compounds are used to couple the fundamental energy sources to the energy-requiring processes of the cell.

8 The oxidative pathway and oxidative phosphorylation

We have seen in Chapter 7 that oxidative processes are prominent among the energy sources available to the cell. Oxidations of organic compounds in which molecular oxygen is used as the final electron acceptor will yield as much as 200 kJ/g-atom of oxygen. In biological oxidations this yield of free energy is coupled to the synthesis of ATP and is not primarily expressed in the production of heat. We have already seen that the synthesis of ATP from ADP requires about 30 kJ/mol; thus several molecules could result from one, efficiently coupled oxidation. A multistage process known as 'oxidative phosphorylation' exists for this purpose. To carry out the process in stages has the advantage that the energy may be taken off in steps nearer in size to those required for the production of a single molecule of ATP. In fact, three molecules of ATP are produced by the full process, an energy yield of nearly 50%. This compares favourably with the efficiency of the reciprocating steam engine (about 7%) and with the overall efficiency of the steam turbine–dynamo couple in an electricity generating station (about 30%).

Other modes of biological oxidation do exist and some involve energy coupling. However, oxidative phosphorylation is of unique importance. It is a common adjunct to the majority of degradative reaction pathways described in this book and the principal source of the ATP required by most cells.

In the cells of higher organisms oxidative phosphorylation takes place, as was mentioned on p. 00, in an organized structure within the cell known as the mitochondrion (Chapter 21). There are many similarities between the oxidative production and the photosynthetic production of ATP. The latter takes place in an analogous entity in photosynthetic cells, the chloroplast. In both cases the principal catalysts are protein–prosthetic group complexes. Lipids are also present. It is possible that lipids have an active role in the phosphorylation process; they are certainly extremely important in establishing and controlling the critical three-dimensional inter-relationships between the other components.

We will now describe, in a simplified form, the stages of oxidative phosphorylation. Almost certainly much more awaits discovery, but the general outline of our understanding of the process is unlikely to change to a significant extent.

The hydrogen and electron transport pathway

The oxidation of an organic compound may be written

$$AH_2 + \tfrac{1}{2}O_2 \rightleftharpoons A + H_2O. \qquad (1)$$

For the majority of compounds the first stage, as we saw in Chapter 7, is

$$AH_2 + NAD(\text{oxidized}) \rightleftharpoons A + NAD(\text{reduced}). \qquad (2)$$

This will be an enzyme-catalysed reaction. The enzyme may either be absolutely specific for A or catalyse the dehydrogenation of a range of similar compounds.

NAD (p. 104) is usually bound (non-covalently) to the enzyme that catalyses reaction (2). The strength of the binding is more characteristic of a co-factor than a prosthetic group. The part of the molecule which is concerned in reduction–oxidation is the nicotinamide ring (Fig. 8.1). (It should be possible to see from this figure why the oxidized form of NAD is often written as NAD^+ and the reduced form NADH, with a similar notation for NADP. We prefer the notation $NADH_2$ for the reduced form and NAD for the oxidized form.)

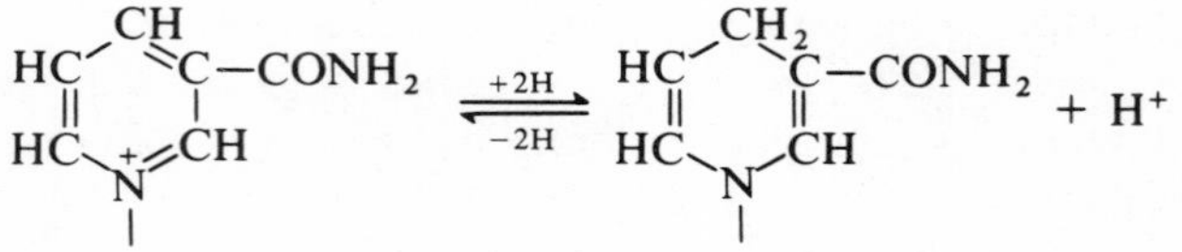

Fig. 8.1. Oxidation and reduction in the nicotinamide ring.

It is beyond the capacity of human metabolism to synthesize nicotinamide. The vital place of NAD^+ and NADP in metabolism explains why nicotinamide is a vitamin (see also p. 29).

The next stage in the transport of hydrogen from the compound AH_2 is the reduction by $NADH_2$ of another ring system (Fig. 8.4). The ring is found in nature as part of two molecules, flavin mononucleotide (FMN) and flavin adenine dinucleotide (FAD) (see Table 3.2). FMN and FAD are attached to proteins by non-covalent forces; the attachment is strong enough for them to be termed prosthetic groups. The FMN– and FAD–protein complexes are known as flavoproteins.

Like nicotinamide, riboflavin cannot be synthesized by man and is a vitamin.

The flavoprotein involved in the reaction

$$NADH_2 + \text{flavoprotein(oxidized)} \rightleftharpoons$$

$$NAD + \text{flavoprotein(reduced)} \quad (3)$$

can be thought of as an enzyme catalysing the dehydrogenation of $NADH_2$. Some flavoproteins are, in fact, dehydrogenase enzymes in the more usual sense – for example, in the desaturation of aliphatic carbon

$$R_1CH_2CH_2R_2 + \text{flavoprotein(oxidized)} \rightleftharpoons$$

$$R_1CH{=}CHR_2 + \text{flavoprotein(reduced)}$$

(see pp. 137 and 141). Such dehydrogenations therefore bypass NAD, but from then on they follow the same pathway as the rest.

We have seen that the early stages of the pathway summarized in equation (1) involve the transfer of *hydrogen atoms* between complex ring molecules bound to proteins. If we turn to the final stages we find that these consist of *electron* transfer down a chain of protein–prosthetic group complexes (the cytochromes, see below), ultimately ending in oxygen.

The cytochromes rely on ferro-porphyrin rings similar to those used by haemoglobin. The three main families of cytochromes

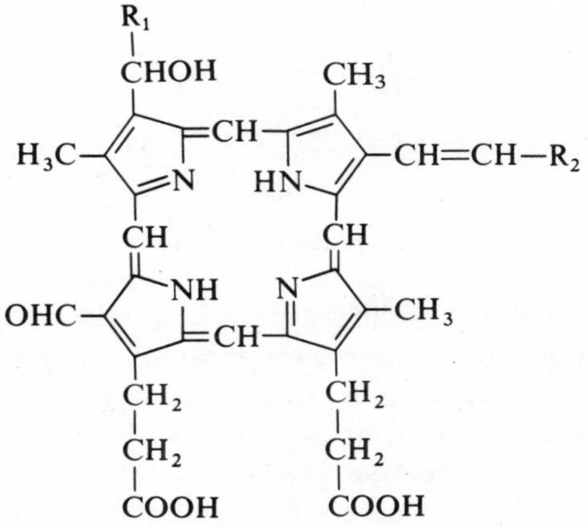

cytochrome *a*

Fig. 8.2. Substitutions in the porphyrin rings of the cytochromes.

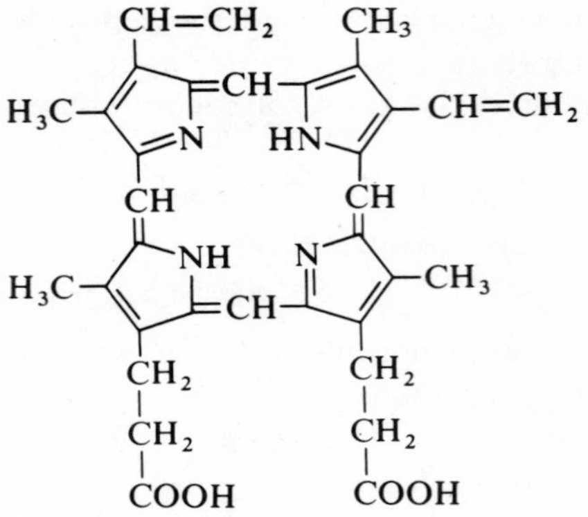

cytochrome *b*

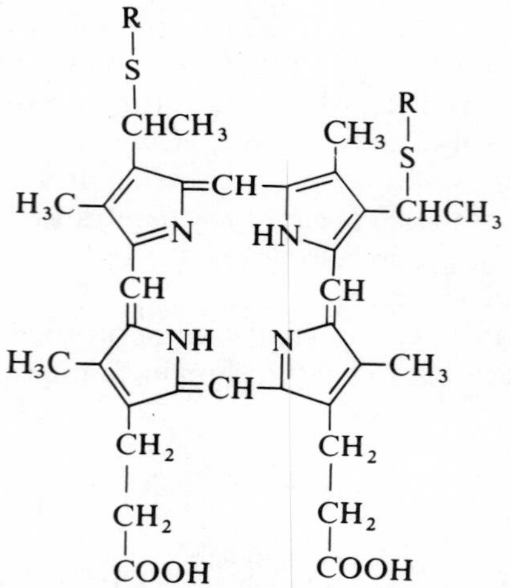

cytochrome *c*

Fig. 8.2.—*continued*

(cytochromes *a*, *b* and *c*) differ from one another by virtue of their selection of R groups at the corners of the ring (Fig. 8.2). Individual members of a family differ from one another in their amino-acid sequence and not necessarily in the ring substitution.

Cytochrome *c* is the best studied of the cytochromes and is probably typical. In distinction to haemoglobin (p. 40) the ring is covalently as well as non-covalently bound to the protein. The sixth co-ordination position of the iron, which in haemoglobin was free to receive oxygen, is permanently occupied by an amino-acid side chain. Rather than being exposed on the surface of the protein, the ring is buried in a crevice and only one edge lies on the surface (Fig. 8.3). This serves to emphasize that the cytochromes act as conduc-

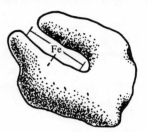

Fig. 8.3. Cytochrome *c*.

tors of electrons and that their primary mode of action does not, in distinction to the case of haemoglobin, involve the binding of ions or molecules to the ring.

How the two parts of the oxidative pathway, hydrogen transfer and electron transfer, are linked is not quite clear. Almost certainly another ring compound, ubiquinone, accepts hydrogens from the isoalloxazine ring of the flavoprotein (Fig. 8.4). The long hydrocarbon chain of ubiquinone would favour hydrophobic binding to the lipid of the mitochondrion. This is the last easily recognized

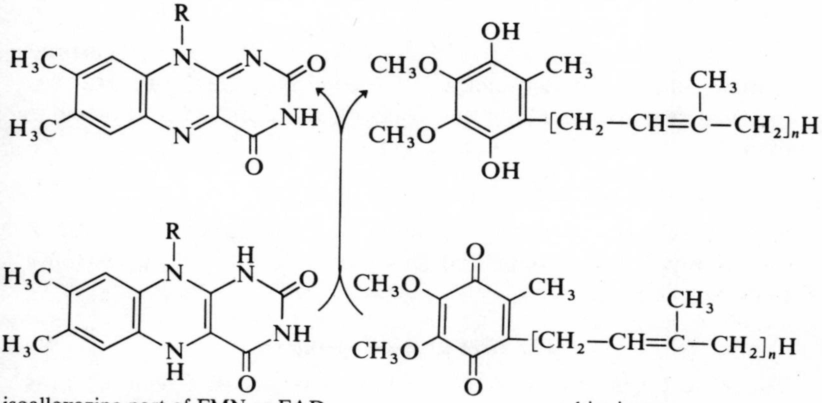

isoalloxazine part of FMN or FAD   ubiquinone

(For the nature of R see Table 3.2)

Fig. 8.4.

transfer of hydrogen atoms; thereafter a step occurs which may formally be written as:

$$2H \text{ (from the hydrogen transfer pathway)} \rightleftharpoons 2H^+ + 2e^- \quad (4)$$

and electron transfer commences. The electrons are handed on, one at a time, from cytochrome to cytochrome. The iron atom in the ferro-porphyrin ring of each shuttles backwards and forwards between Fe^{3+} and Fe^{2+} as each electron is passed on. (This is a further point of difference between the cytochromes and haemoglobin, since the iron in the latter remains in the Fe^{2+} state whether or not a molecule of O_2 is attached to it.)

We have already seen that the whole process is taking place in a highly unusual environment, a lipid–protein matrix. Therefore, while equation (4) and equation (5) below are useful for a general understanding of what takes place, we need not necessarily assume that the protons and electrons ever exist as identifiable separate entities. Also, the succession of reactions is so fast that once again (see p. 39) we are led to assume that there is no free diffusion of intermediates from one protein to another. The active part of the carriers must be almost in contact and the product of one reaction handed directly on to the catalyst of the next.

How reaction (4) is brought about is far from clear except that other, metal-containing, proteins appear to be involved. Thereafter, matters appear to be straightforward with the transfer of electrons from one cytochrome to another. The order is given in Fig. 8.5.

Finally cytochrome a or a_3 (the separate existence of cytochrome a and a_3 has been questioned), with the help of a Cu^{2+} ion, catalyses the final reaction in which the electrons are donated to molecular oxygen:

$$2e^- + 2H^+ + \tfrac{1}{2}O_2 \rightleftharpoons H_2O \quad (5)$$

and the oxidative pathway is at an end. (Cyanide is an inhibitor of the terminal cytochrome and this fact accounts for its extreme toxicity.)

The sites of ATP synthesis

Those oxidations catalysed by enzymes that are also flavoproteins yield only two molecules of ATP per atom of oxygen used. Those involving NAD-linked dehydrogenases yield three, as does the oxidation of $NADH_2$ itself. Therefore, one phosphorylation must be associated with equation (3). Of the other two sites, one appears

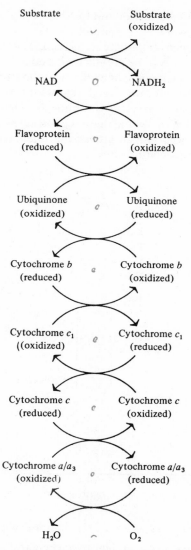

Fig. 8.5. The hydrogen and electron transport pathway.

to be between the ubiquinone/cytochrome b region and cytochrome c_1, the other on the oxygen side of cytochrome c. We have
several lines of evidence for these conclusions, including estimates
of the energetics of the individual reactions. These are, however,
difficult to determine. The normal electrode potential and $\Delta G^{0\prime}$

determinations can be carried out for the molecules involved when in free solution: as we have pointed out more than once these are not the circumstances obtaining in the cell. The values obtained in free solution are likely to be greatly modified in the 'solid state', but nonetheless some estimate can be made of the true situation. Thus, while somewhat speculative, Fig. 8.6 shows that some steps do have the energy span necessary for the phosphorylation, while others certainly do not. On the whole there is good agreement between the energetic calculations and all other experimental data on the location of the phosphorylation sites.

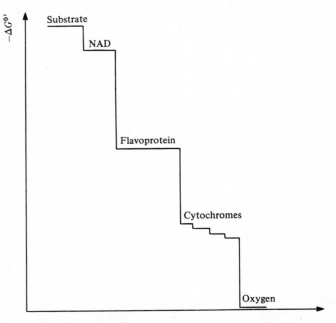

Fig. 8.6. Energy changes in hydrogen and electron transport.

The mechanism of the phosphorylation

The transfer of hydrogen, and more particularly electrons, seems at first sight to have little to do with the reaction:

$$ADP + phosphate \rightleftharpoons ATP + H_2O.$$

In fact all chemical reactions are electron processes and we should not find the connection too hard to imagine. Imagine is the word, unfortunately, because in spite of much experimental work little has emerged that is final. Of the possible mechanisms that have been suggested, we shall describe in outline only the two most favoured possibilities.

The first of these possible explanations is known as the *chemical-coupling* hypothesis. We mentioned on p. 103 that, in addition to the process of *oxidative* phosphorylation which we are now considering, some phosphorylations of ADP take place that are not at all connected with the electron transport pathway but are part of the mechanism of straightforward chemical reactions. We shall give an example of these *substrate-level* phosphorylations (glance for a moment at p. 129), in which the complete chemical process appears to be well understood. It is therefore tempting to look for similar chemical transformations of intermediate compounds which might be associated with the hydrogen and electron transport pathway.

In the example of substrate-level phosphorylation given on p. 136, the oxidation of a molecule that is not a high-energy compound (α-oxoglutaric acid) produces one (a thioester), which, as we saw in Chapter 7, is a high-energy compound. Could not similar transformations be taking place in the mitochondrion, with the original low-energy compound being regenerated after the high-energy bond is used to make ATP?

That is,

$$C_{red} + I \rightleftharpoons C_{ox} + I^*$$
$$I^* + ADP + P_i \rightleftharpoons ATP + I$$

where C is the appropriate carrier in the transport chain and I is the intermediate compound (written as I^* when in its activated state). I is called the *auxiliary coupling factor*.

This states the matter at its simplest. After much work it is clear that if the hypothesis is to be tenable at all the scheme must be at least as complex as

$$C_{red} + I + C'_{ox} \rightleftharpoons C_{ox} \sim I + C'_{red}$$
$$C_{ox} \sim I + enzyme \rightleftharpoons C_{ox} + enzyme \sim I$$
$$enzyme \sim I + P_i \rightleftharpoons I + enzyme \sim P$$
$$enzyme \sim P + ADP \rightleftharpoons ATP + enzyme$$

where $\sim$ indicates a high-energy bond between two substances.

That is, this hypothesis would state that as the reduced carrier, C_{red}, is oxidized (by the next carrier C') it forms a high-energy chemical bond with I. I is then transferred to an enzyme, with retention of the high-energy bond. This is the enzyme that forms ATP and it does so by first exchanging the high-energy bond to I for a high-energy phosphate and then donating this phosphate group to ADP.

Quite a large amount of evidence may be reconciled with the model. However, the most intensive search over some years has failed to detect any of the high-energy intermediates ($\sim$I, $\sim$P compounds) in a direct experiment. This failure may be due to the necessity, in a system bringing about oxidative phosphorylation, for a high degree of structural organization, which tends to be lost in the kind of test-tube experiment that would be necessary to isolate $\sim$I compounds. Experiments on fragments of mitochondria show that the system does indeed rely on a measure of organization, but to so great a degree as to give rise to the suspicion that there is more to the story than we have so far considered. An emphasis on structural organization leads, in fact, to an entirely different hypothesis, which we shall now examine.

In order to appreciate this alternative, the *chemi-osmotic* hypothesis, we must know that the mitochondrion is bounded by a double membrane, and that the outer membrane is very different from the inner one in composition and properties. The outer membrane is richer in lipid and poorer in protein than most membranes, and is simpler in terms of the functions that it undertakes. It allows the passive entry of a larger number of solutes of small molecular weight and, mirroring its lower protein content, it has few specific types of enzymic activity associated with it. The inner membrane by contrast has a number of systems for the transport of certain specific substances (cf. p. 219). Moreover, it is in the inner membrane that the carrier system for the oxidative pathway is located. (Many other integrated, multiprotein systems are also found here.) The regular disposition, all over the inner membrane, of assemblies of the protein carriers is striking. Now it is found that if the inner membrane is broken up, phosphorylation is no longer possible *unless the pieces are able to re-form themselves into closed bodies, entirely surrounded by membrane.*

It is this last observation that provides powerful support for the chemi-osmotic hypothesis. Briefly stated, the hypothesis is as follows:

As reducing power and electrons are transported down the oxidation chain, H^+ ions are expelled from the outer surface of the inner membrane. Six protons are expelled for the passage of the equivalent of two electrons; these protons accumulate outside the membrane, so that the lowered pH on the outside is balanced by a raised pH (an excess of OH^-) on the inside. To concentrate a substance requires free energy; the free energy of the fall of the electrons down the chain is therefore, if this model is correct, stored in the form of the concentration gradient of H^+ that the expulsion creates.

One possible mechanism by which this gradient of $[H^+]$ could be set up would be that the steps of the transport pathway that involve H^+ (the first three and then the terminal step) are catalysed by carriers that are so oriented that they can *take up* protons only from the *inside* of the membrane and *discharge* them only to the *outside*. All that we know of the structures of proteins and of membranes makes it easy for us to believe that this is quite possible. The appropriate determinations of pH are extremely difficult, but such results as have been obtained appear to confirm that a detectable concentration gradient of H^+ can indeed be set up. (The energy inherent in the concentration gradient may be partly converted to that inherent in a gradient of electric *charge*. We have already emphasized that a difference in electrical potential is equivalent, in terms of free energy, to a difference in concentration.)

If we accept for the moment that the energy of the fall of electrons to oxygen is first stored as a concentration gradient, or as an equivalent gradient of charge, how is this then used for the phosphorylation of ADP?

The phosphorylation of ADP,

$$ADP + P_i \rightleftharpoons ATP + H_2O,$$

involves the condensation of the P_i to form the high-energy phosphoric anhydride bond with the exclusion of water. The equation could be written more fully as

$$\text{Adenosine}-\overset{\displaystyle O}{\overset{\|}{\underset{\underset{\displaystyle OH}{|}}{P}}}-O-\overset{\displaystyle O}{\overset{\|}{\underset{\underset{\displaystyle OH}{|}}{P}}}-OH \;+\; HO-\overset{\displaystyle O}{\overset{\|}{\underset{\underset{\displaystyle OH}{|}}{P}}}-OH \;\rightleftharpoons$$

$$\text{Adenosine}-\overset{\displaystyle O}{\overset{\|}{\underset{\underset{\displaystyle OH}{|}}{P}}}-O-\overset{\displaystyle O}{\overset{\|}{\underset{\underset{\displaystyle OH}{|}}{P}}}-O-\overset{\displaystyle O}{\overset{\|}{\underset{\underset{\displaystyle OH}{|}}{P}}}-OH \;+\; H_2O \qquad (6)$$

but for the various reasons that we gave in our note on abbreviations and conventions (p. xi) this is not normally done.

The $\Delta G^{0\prime}$ of this reaction is such as to produce an equilibrium lying very far over to the left; in a normal environment we could not expect to be able either to push or pull the reaction to the right simply by the manipulation of concentrations (p. 97). But the highly organized hydrophobic environment of the mitochondrial membrane is not 'normal' in this sense, and there is some evidence to suggest that, just as the chemi-osmotic hypothesis demands, the membrane system does indeed pull the reaction over by making use of energy stored in the $[H^+]$ and electrical gradient. The energetics of such a process are not at all unreasonable: there could be sufficient energy stored in the concentration and charge gradients to pull the reaction over to net synthesis of ATP.

All this might seem far-fetched in comparison with the chemical-coupling hypothesis but, if it does, it is probably because we bring prejudices to the subject that we have acquired from the study of non-living matter and even from the study of the simpler bio-chemical processes. It is easy to feel that a set of simple, definable chemical reactions is more plausible than an appeal to seemingly vague directional properties of a membrane. There is, however, abundant indication that the living cell is fully capable of produc-ing such directional components if necessary. The need for an intact membrane around any structure that is to carry out oxidative phosphorylation, and the failure to isolate the ∼I or ∼P inter-mediates, do nothing to enhance our faith in the 'straightforward' nature of the process. Concentration (or charge) gradients of the type postulated are often observed, and much of the evidence fits the hypothesis very well. For instance, a large protein complex has been found on the inner membrane which, when isolated, catalyses the hydrolysis of ATP, and when in the intact system is able to promote the specific transport of protons across the membrane.

A final adjudication between the two types of hypothesis has yet to be made, but opinion seems to be progressively moving in favour of some kind of chemi-osmotic theory.

Leaving now the details of the coupling, we shall close this chapter by repeating the main facts, none of which are seriously in dispute. These are that:

(i) the vital task of trapping the free energy generated by the fall of electrons to molecular oxygen is undertaken by a tightly integrated chain of carriers;

(ii) the first clearly recognizable chemical form in which the energy resides is ATP, the major energy carrier of the cell;

(iii) the number of molecules of ATP produced from the oxida-tion of an amount of substrate possessing the equivalent of 2H is most probably three, an overall efficiency of about 50%.

9 Introduction to intermediary metabolism

We have now set the scene for a discussion of some aspects of *intermediary metabolism*. This name is given to that set of reactions by which the small molecules found in biological systems are degraded, synthesized, interconverted and otherwise chemically transformed. A striking characteristic of these transformations is that they occur in very small, discrete steps, each of which is catalysed by a separate enzyme. Thus intermediary metabolism consists of a large number of reactions, by means of which molecules are gradually modified and shaped. The various reactions are generally thought of as comprising 'pathways', along which the compounds flow while undergoing these gradual changes. However, the pathways constantly converge and diverge, so that an intermediate formed as a result of one reaction may often have a choice of two or more subsequent paths to follow.

One reason why metabolic transformations involve so many steps will become clear if we consider the mode of action of enzymes. If metabolism is to be at all vigorous, it is often necessary for enzymes to be extremely active. In order to increase the rate of a biochemical reaction the enzyme must bind to the substrate and to the reaction intermediate (p. 47). For the increase in rate to be appreciable, the substrate and subsequently the reaction intermediate must be very strongly bound through a number of firm, specific contacts with the enzyme. It is remarkable that a linear polymer of amino acids can do so well in providing a binding surface even for normal enzyme-catalysed reactions, in which only a few atoms are involved: to provide a surface that could organize (for example) the simultaneous oxidation of all six carbon atoms of glucose would be considerably more difficult.

Thus as a consequence of the need of enzymes to possess a catalytic power that is, in many instances, immense, the chemical change that any one enzyme reaction brings about tends to be correspondingly slight. This may well be an important reason for the fact that the pathways of intermediary metabolism involve such a gradual modification of the reacting molecules.

The fact that any one enzyme does such a specific, small-scale job necessitates the presence in organisms of a large number of different enzymes. One might think that, even despite the points that we made above, it would be better for organisms not to have to make so many enzymes. For example, we shall see in Chapter 10 that the breakdown of glucose to lactic acid requires eleven separate enzymes, each of which must be made by the specific protein-synthesizing machinery of the cell (Chapter 20). Would it not be more economical to have a single enzyme that achieves the production of lactic acid from glucose, even if that enzyme had a much more complex structure and, perhaps, less catalytic power than the enzymes that actually exist?

The answer is that this division of metabolic reactions into small steps is admirably suited to fulfilling two separate functions undertaken by intermediary metabolism. These are to provide energy in the form of ATP and to produce the building blocks from which macromolecules are synthesized. In the degradation of glucose to lactic acid several intermediates are formed which are used in other pathways – for example glucose-6-phosphate which can be converted to glycogen (Chapter 15), glyceraldehyde-3-phosphate which can be converted to glycerol and thus used in the synthesis of fats (Chapter 16) and pyruvic acid which (among many other uses) can be converted to alanine and used in the synthesis of proteins (Chapter 20). A similar multiple use is made of every pathway in intermediary metabolism. If glucose were degraded to lactic acid in one step the essential intermediates would have to be formed by other routes, each of which would require its own separate enzyme or enzymes. So the advantage that would seem to result if enzymes were designed to achieve much more radical transformations would in fact be no advantage at all.

On the contrary, there is yet another advantage in splitting up metabolic pathways into small steps. The economy of the cell demands that the processes of intermediary metabolism be carefully controlled, so that the ATP and intermediates that they provide should be available in the right quantities at the right time. Now by dividing metabolic pathways into a large number of steps the cell can effect a very fine control of pathways that deal with *several* different substrates by altering the rate of just a single reaction. For example, we shall see that the degradation of glucose to lactic acid involves a reaction catalysed by the enzyme phosphofructokinase. This enzyme, however, is needed equally for the degradation of lactose via galactose and of glycogen, in each case to lactic acid. So by blocking the activity of phosphofructokinase the cell can block the degradation of all these carbohydrates (p. 239) without at the same time affecting any other metabolic pathway.

Although other general points about intermediary metabolism could be made, we have said enough to explain some of the reasons for what might seem a bewildering complexity of metabolic reactions. In the chapters that go into intermediary metabolism in some detail we have been obliged to separate, for the purposes of exposition, the two functions of the reactions that we describe – the formation of ATP and the production of useful intermediates for biosynthesis. But it is important not to lose sight of their interrelationship, and in order to give an overview of the subject we shall now summarize the contents of Chapters 10–17, so that each pathway can be seen in its context.

If we first consider the chief energy-yielding pathways, we can

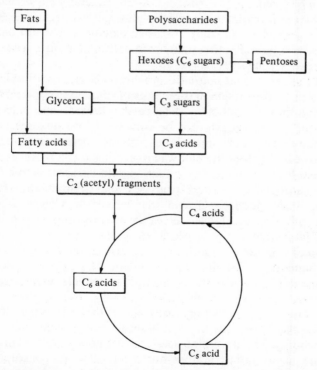

Fig. 9.1. Sketch-map of metabolic reactions to show the degradation of carbohydrates and fats.

draw a chart (Fig. 9.1) that gives the outline of the processes. Polysaccharides are important storage molecules (Chapter 6); and these are depolymerized to hexoses (sugars that contain six carbon atoms). Hexoses are broken down via C_3 sugars (sugars that contain three carbon atoms) to C_3 acids, as we shall see in Chapter 10. C_3 acids are oxidized to give acetyl (C_2) fragments, and these join with a C_4 acid to give a C_6 acid, which is then successively oxidized via a C_5 acid to yield a C_4 acid again which can once more accept an acetyl fragment (Chapter 11); this sequence of reactions thus completely oxidizes the C_3 acids, and hence the hexoses. The other important storage molecules are fats, which are composed of glycerol and fatty acids. Glycerol is converted to a C_3 sugar; fatty acids are oxidized to acetyl fragments which are degraded as before (Chapter 12).

During the course of these degradations, pairs of hydrogen atoms are removed from the intermediates at several stages in the pathways. These pairs of hydrogen atoms are generally passed to NAD and thus to the oxidative phosphorylation pathway (Chapter 8), and the synthesis of ATP results from their oxidation. The one pathway that reduces NADP rather than NAD is the oxidation of a hexose to a pentose (C_5 sugar) (Chapter 13).

By contrast, photosynthetic organisms derive both ATP and $NADPH_2$ by conversion of the energy of sunlight (Chapter 14), and they use these compounds to fix carbon dioxide into a C_3 sugar (Chapter 15). C_3 sugars can be converted into hexoses both in photosynthetic and non-photosynthetic organisms; but (as indicated in Fig. 9.2) the pathway of synthesis of hexoses is in part different from the path of degradation, and the same is true for the synthesis of polysaccharides from hexoses (Chapter 15). Hexoses can also be formed from C_3 acids that derive from C_4 acids (Chapter 15). In addition, as Fig. 9.2 shows, fatty acids can be formed from acetyl fragments; but the pathway of synthesis is different from the degradative pathway, and this is true also of the formation of the fats themselves. Equally the conversion of pentoses to hexoses (Chapter 13) is not just the reverse of the conversion of hexoses to pentoses.

We can now see how the intermediates that are formed in these pathways are used in the synthesis of the building blocks for macromolecules – the amino acids that are constituents of proteins, and the purine and pyrimidine nucleotides that are constituents of nucleic acids. Fig. 9.3, which is an expansion of the left-hand side of Fig. 9.1 and Fig. 9.2, shows some of these synthetic pathways.

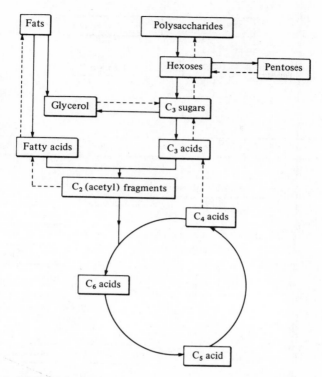

Fig. 9.2. Sketch-map of metabolic reactions to show the degradation and synthesis of carbohydrates and fats.

Amino acids can be made from several of the compounds that are produced from the breakdown of carbohydrates, such as C_3 acids and, more important, C_4 and C_5 acids (Chapter 17). The C_4 amino acid, aspartic acid, gives rise to other amino acids (Chapter 17) and also to pyrimidines (Chapter 19); the C_5 amino acid, glutamic acid, gives rise to still other amino acids (Chapter 17). The skeleton of purines (unlike that of pyrimidines) is formed piecemeal from a number of precursors (Chapter 19). The ribose phosphate and deoxyribose phosphate that are constituents of nucleic acids are formed from the pentose produced by oxidation of hexose (Chapter 13).

The next seven chapters describe in much more detail the pathways that we have mentioned here. You will find that these chapters are extensively cross-referenced: it is our intention to stress the interrelationships of the metabolic pathways that we discuss, and

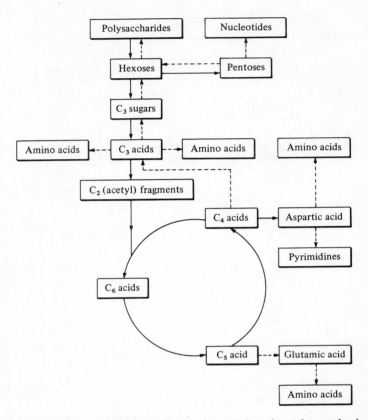

Fig. 9.3. Sketch-map of metabolic reactions to show how the synthesis of amino acids and nucleotides is related to intermediates in the degradation of carbohydrates.

thus to show how intermediary metabolism combines the two functions that we have mentioned – providing energy and supplying the intermediates needed for the synthesis of macromolecules.

10 Synthesis of ATP – glycolysis

Glucose is the key sugar in carbohydrate metabolism. We have previously seen (Chapter 6) that starch, the polysaccharide storage material of plants, and glycogen, the polysaccharide storage material of animals, both consist of glucose residues. Moreover, other sugars can be converted to glucose by simple series of reactions. Again, carbohydrate is supplied in the form of glucose to most cells of the animal body by the blood. Thus in discussing the breakdown of carbohydrates we shall first consider the breakdown of glucose and then show how some other compounds fit into the scheme.

Organisms that live under anaerobic conditions (that is to say, in the absence of oxygen) can break down glucose by *fermentation* reactions. In the course of fermentation the molecule is split without undergoing either net oxidation or net reduction, so that the number of hydrogen and oxygen atoms in the products is the same as in the starting material. At first sight fermentations appear to be very varied, because different organisms accumulate different products of fermentation – lactic acid, ethanol, butanol, acetone, propionic acid, etc., etc. But this apparent diversity conceals the fact that most of the reactions in the pathway are actually the same for different organisms; it is only at the very last stages that the pathways in different species diverge. So we are justified in looking, to begin with, at just one fermentation as an example of the anaerobic breakdown of glucose. The most convenient one to examine is the formation of lactic acid.

Although the fermentation of glucose to produce lactic acid is an anaerobic process, it is also of great importance to organisms that make use of molecular oxygen to oxidize carbohydrates. There are two reasons why aerobic organisms have adopted the reactions of the fermentation pathway in order to break down glucose. First, even in aerobic organisms like ourselves, anaerobic conditions sometimes prevail temporarily in the muscles, and lactic acid appears in quite large quantities in actively working muscle. Secondly, even in tissues that never become anaerobic much of the

125

metabolism of glucose still involves borrowing the reactions of the fermentation pathway. These reactions split the glucose molecule in two, and the reactions of the aerobic pathway (see next chapter) then oxidize the fragments thus formed.

Glycolysis

The anaerobic breakdown of glucose is often called *glycolysis*, and it proceeds via a pathway often called the Embden–Meyerhof pathway. Although the detailed reactions of the Embden–Meyerhof pathway are quite complicated, the net result is extremely simple. 'Glycolysis' is the splitting of glucose, and the sum total of the reactions is just this:

$$C_6H_{12}O_6 \rightarrow 2C_3H_6O_3,$$

$$\text{glucose} \rightarrow \text{lactic acid.}$$

It is evident that this reaction involves no net oxidation or reduction, and is therefore a true fermentation.

Let us now look at the details of this process. Before glucose can undergo any metabolic change, it must first be phosphorylated by an enzyme called hexokinase. The equilibrium constant of this reaction overwhelmingly favours the formation of glucose-6-phosphate.

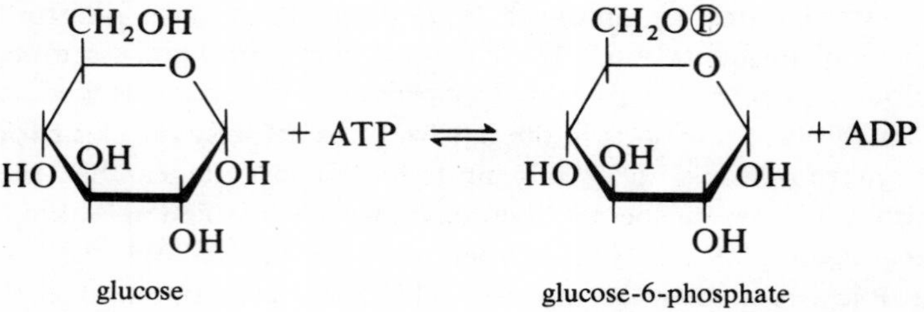

glucose   glucose-6-phosphate

Glucose-6-phosphate is converted in the cell to fructose-6-phosphate, the reaction being catalysed by the enzyme hexose phosphate isomerase. Fructose-6-phosphate can now be phosphorylated once more at the expense of ATP. The reaction is catalysed by phosphofructokinase and yields fructose-1,6-bisphosphate. As with the phosphorylation of glucose the $\Delta G^{0\prime}$ greatly favours the formation of the product.

Fructose-1,6-bisphosphate can be cleaved by the action of the enzyme aldolase to give two molecules of triose phosphate. One of

glucose-6-phosphate   fructose-6-phosphate   (2)

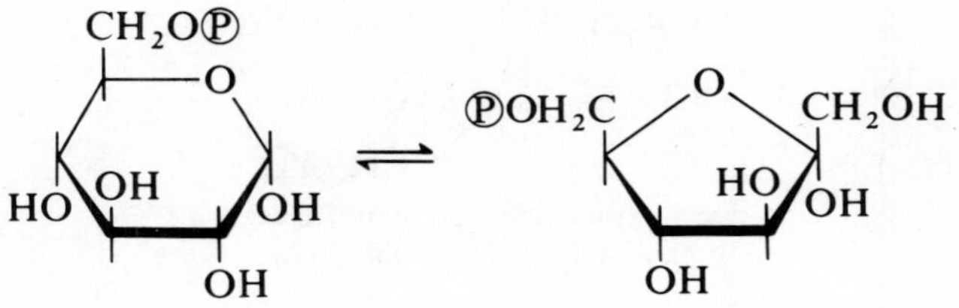

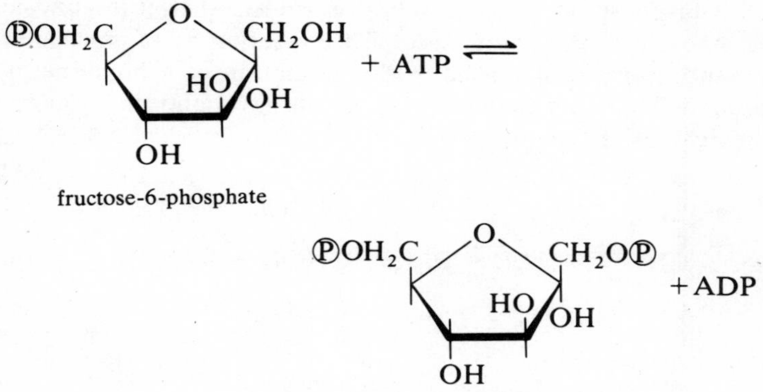

fructose-6-phosphate

$+ADP$ (3)

fructose-1,6-bisphosphate

these is dihydroxyacetone phosphate and the other glyceraldehyde-3-phosphate. These two molecules are themselves easily interconverted in the cell. The interconversion is catalysed by the enzyme triose phosphate isomerase. As we have already discussed (p. 96) the next reaction in the glycolytic sequence uses only glyceraldehyde-3-phosphate, but because of the reaction just mentioned dihydroxyacetone phosphate is converted into glyceraldehyde-3-phosphate as the latter is used up. Thus *both* of the triose phosphates derived from fructose-1,6-bisphosphate are in fact further metabolized.

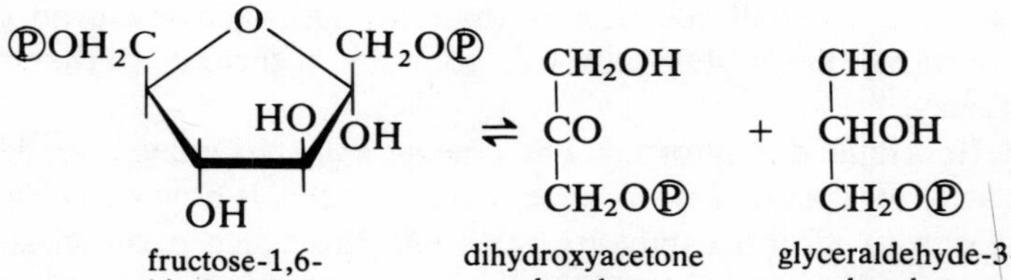

fructose-1,6-bisphosphate   dihydroxyacetone phosphate   glyceraldehyde-3-phosphate   (4)

$$
\begin{array}{ccc}
\begin{array}{c}
\text{CH}_2\text{OH} \\
| \\
\text{CO} \\
| \\
\text{CH}_2\text{O}\circled{P} \\
\text{dihydroxyacetone} \\
\text{phosphate}
\end{array}
& \rightleftharpoons &
\begin{array}{c}
\text{CHO} \\
| \\
\text{CHOH} \\
| \\
\text{CH}_2\text{O}\circled{P} \\
\text{glyceraldehyde-3-} \\
\text{phosphate}
\end{array}
\end{array}
\qquad (5)
$$

We have now arrived at a most significant reaction of glycolysis, that in which glyceraldehyde-3-phosphate is oxidized to glyceric acid-1,3-bisphosphate. The reaction is catalysed by the enzyme triose phosphate dehydrogenase. This reaction is of particular importance because it results in the formation of a high-energy compound. This is an example of substrate-level phosphorylation, which yields ATP (see p. 103)

$$
\begin{array}{ccc}
\begin{array}{c}
\text{CHO} \\
| \\
\text{CHOH} + \text{NAD} + \text{P}_i \\
| \\
\text{CH}_2\text{O}\circled{P} \\
\text{glyceraldehyde-3-} \\
\text{phosphate}
\end{array}
& \rightleftharpoons &
\begin{array}{c}
\text{COO}\circled{P} \\
| \\
\text{CHOH} + \text{NADH}_2 \\
| \\
\text{CH}_2\text{O}\circled{P} \\
\text{glyceric acid-1,3-} \\
\text{bisphosphate}
\end{array}
\end{array}
\qquad (6)
$$

$$
\begin{array}{ccc}
\begin{array}{c}
\text{COO}\circled{P} \\
| \\
\text{CHOH} + \text{ADP} \\
| \\
\text{CH}_2\text{O}\circled{P} \\
\text{glyceric acid-1,3-} \\
\text{bisphosphate}
\end{array}
& \rightleftharpoons &
\begin{array}{c}
\text{COOH} \\
| \\
\text{CHOH} + \text{ATP} \\
| \\
\text{CH}_2\text{O}\circled{P} \\
\text{glyceric acid-3-} \\
\text{phosphate}
\end{array}
\end{array}
\qquad (7)
$$

At first sight the oxidation of glyceraldehyde-3-phosphate and the accompanying reduction of NAD appear to conflict with the principle that we mentioned earlier – that glycolysis does not involve net oxidation or reduction of the glucose molecule. However, we shall see shortly that the hydrogen is stored only temporarily as $NADH_2$; it will soon be returned to the main pathway.

Glyceric acid-3-phosphate is changed (by an enzyme called a mutase) to glyceric acid-2-phosphate, and this is now dehydrated. The dehydration is catalysed by an enzyme called enolase, and it results in the formation of the high-energy compound phospho-

*enol*pyruvic acid (see p. 103).

$$
\begin{array}{ccc}
\text{COOH} & & \text{COOH} \\
| & & | \\
\text{CHOH} & \rightleftharpoons & \text{CHO}\textcircled{P} \\
| & & | \\
\text{CH}_2\text{O}\textcircled{P} & & \text{CH}_2\text{OH}
\end{array} \quad (8)
$$

glyceric acid-3-   glyceric acid-2-
phosphate   phosphate

$$
\begin{array}{ccc}
\text{COOH} & & \text{COOH} \\
| & & | \\
\text{CHO}\textcircled{P} & \rightleftharpoons & \text{CO}\textcircled{P} \;+\; \text{H}_2\text{O} \\
| & & || \\
\text{CH}_2\text{OH} & & \text{CH}_2
\end{array} \quad (9)
$$

glyceric acid-2-   phospho*enol*pyruvic
phosphate   acid

As we explained on p. 98 the free energy of hydrolysis of phospho-*enol*pyruvic acid is very high, and the enzyme pyruvate kinase can readily catalyse the synthesis of ATP. (We mentioned that the free energy of hydrolysis of phospho-*enol*pyruvic acid is so exceptionally high that the reaction is not reversible to any great extent.) The removal of the phosphate group from phospho-*enol*-pyruvic acid leaves pyruvic acid.

$$
\begin{array}{ccc}
\text{COOH} & & \text{COOH} \\
| & & | \\
\text{CO}\textcircled{P} \;+\; \text{ADP} & \rightleftharpoons & \text{CO} \;+\; \text{ATP} \\
|| & & | \\
\text{CH}_2 & & \text{CH}_3
\end{array} \quad (10)
$$

phospho-*enol*pyruvic   pyruvic
acid   acid

The glycolytic pathway is nearly complete – nearly, but not quite. Reaction (6) of the pathway reduced NAD to $NADH_2$; and since the concentration of NAD in cells is very low it is evident that glycolysis could not continue for any length of time without regeneration of NAD to act once again as a hydrogen carrier in reaction (6). Under aerobic conditions it would be possible to reoxidize $NADH_2$ by the respiratory chain (p. 108). However, glycolysis is an *anaerobic* pathway, and is the principal means of ATP synthesis for many organisms that do not use molecular oxygen for oxidation. How then can NAD be regenerated from $NADH_2$ anaerobically?

The answer is that $NADH_2$ is reoxidized by being used to reduce the product of the glycolytic pathway itself. Pyruvic acid was formed in reaction (10), and the enzyme lactate dehydrogenase readily catalyses the reduction of pyruvic acid to lactic acid.

$$CH_3COCOOH + NADH_2 \rightleftharpoons CH_3CHOHCOOH + NAD \quad (11)$$

pyruvic acid   lactic acid

The NAD can now be used again in reaction (6), and in this way, by recycling NAD, a cell can break down large quantities of glucose and accumulate lactic acid.

Some organisms accumulate not lactic acid but other products of fermentation. This fact need not surprise us if we remember that the function of reaction (11) is to dispose of excess hydrogen and to reoxidize $NADH_2$. Thus any compound that can accept hydrogen from $NADH_2$ can act in place of the pyruvic acid in reaction (11). For example yeast decarboxylates the pyruvic acid first and then uses the product of decarboxylation to reoxidize $NADH_2$.

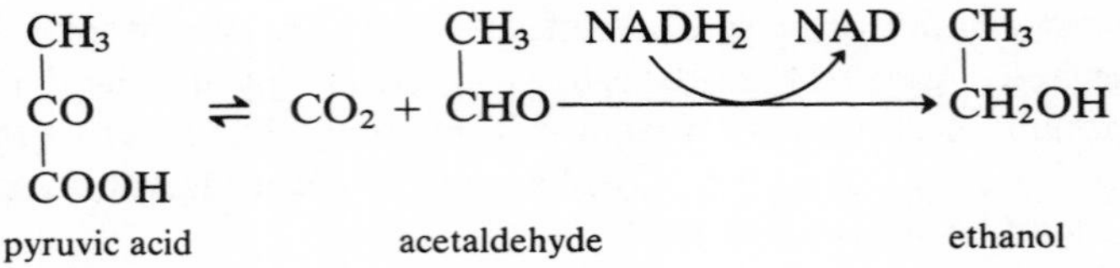

$$\begin{array}{c} CH_3 \\ | \\ CO \\ | \\ COOH \end{array} \rightleftharpoons CO_2 + \begin{array}{c} CH_3 \\ | \\ CHO \end{array} \xrightarrow{\ \ NADH_2 \quad NAD\ \ } \begin{array}{c} CH_3 \\ | \\ CH_2OH \end{array} \quad (12)$$

pyruvic acid   acetaldehyde   ethanol

Other organisms carry out more complex reactions which produce a variety of reduced products, but the biological function is the same in each case.

Many organisms that live anaerobically derive all of their ATP from fermentations of this kind. We can easily calculate the net yield of ATP from glycolysis. One molecule of ATP is used up in each of the reactions (1) and (3). Reaction (4) splits the molecule in halves, and since one ATP is gained in each of the reactions (7) and (10) for each half molecule of glucose the gain from these two reactions is four ATP. Thus the *net* gain of ATP from the fermentation of glucose to lactic acid (or other fermentation products) is two molecules of ATP.

Other compounds and the Embden–Meyerhof pathway

There are several compounds other than glucose that are degraded by the Embden–Meyerhof pathway, although in general two or three special reactions are needed to bring the compound into the pathway. *Fructose*, for example, can be phosphorylated directly to fructose-6-phosphate and enter the pathway at reaction (3).

Galactose is phosphorylated and eventually converted to glucose-6-phosphate. The glycolytic pathway can also be used for the breakdown of *glycerol* in the following way, and the dihydroxyacetone phosphate thus formed can enter at reaction (5).

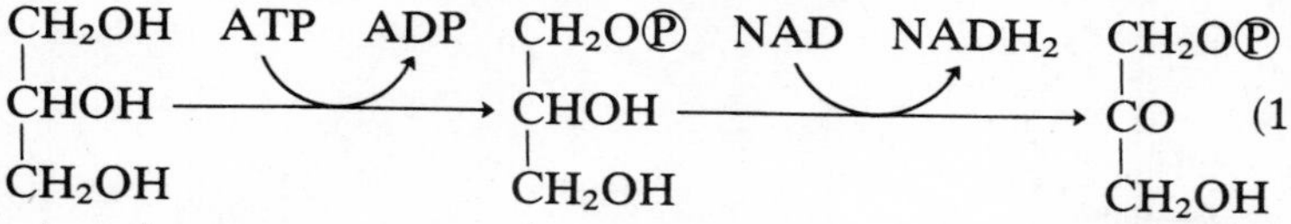

$$
\begin{array}{ccc}
\text{CH}_2\text{OH} & \text{CH}_2\text{O}(P) & \text{CH}_2\text{O}(P) \\
| & \xrightarrow{\text{ATP} \quad \text{ADP}} \quad | & \xrightarrow{\text{NAD} \quad \text{NADH}_2} \quad | \\
\text{CHOH} & \text{CHOH} & \text{CO} \qquad (13) \\
| & | & | \\
\text{CH}_2\text{OH} & \text{CH}_2\text{OH} & \text{CH}_2\text{OH} \\
\text{glycerol} & \begin{array}{c}\text{glycerol-3-}\\\text{phosphate}\end{array} & \begin{array}{c}\text{dihydroxyacetone}\\\text{phosphate}\end{array}
\end{array}
$$

In animals one very important use of the Embden–Meyerhof pathway is to provide energy by the breakdown of glycogen. (In plants starch replaces glycogen.) In animal muscles, particularly, an abrupt demand for ATP is met by glycolysing glycogen without going through glucose as an intermediate. The first step is depolymerization of glycogen by reaction with inorganic phosphate, catalysed by the enzyme phosphorylase.

$$(\text{glucose})_n + \text{P}_i \; \rightleftharpoons \; (\text{glucose})_{n-1} + \text{glucose-1-phosphate} \quad (14)$$

This reaction does not use ATP, so that there is no expenditure of a high-energy compound to bring the glucose residues into the metabolic pathway. (By contrast, phosphorylation of the sugars such as glucose, fructose and galactose, and also of glycerol, *does* require ATP.) A mutase now catalyses the transfer of the phosphate group from the 1 to the 6 carbon atom of glucose (reaction (15)). The glucose-6-phosphate can enter the Embden–Meyerhof pathway at reaction (2).

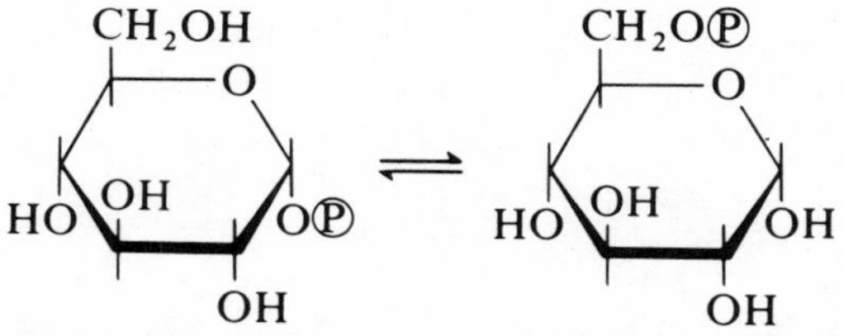

$$\text{(15)}$$

glucose-1-phosphate   glucose-6-phosphate

The animal liver, like the muscle, has a store of glycogen, but its main function here is to replenish glucose in the blood, which is constantly being used by the other tissues. To form glucose from glycogen, the liver first depolymerizes the glycogen with inorganic phosphate, and converts the resulting glucose-1-phosphate to glucose-6-phosphate, by use of reactions (14) and (15). However, the removal of phosphate from glucose-6-phosphate requires a special enzyme, glucose-6-phosphatase, which promotes hydrolysis.

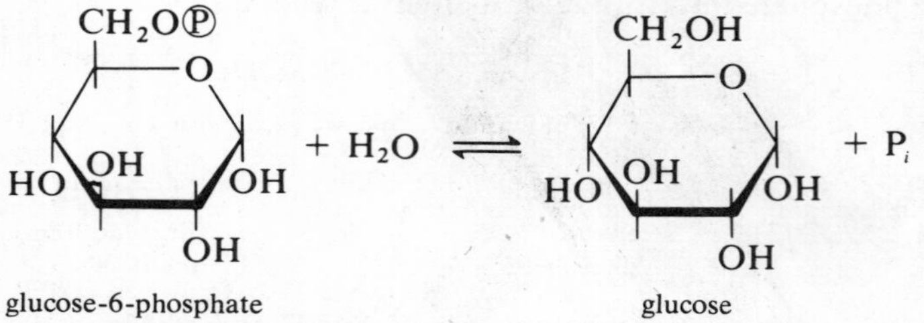

glucose-6-phosphate   glucose

$$\text{glucose-6-phosphate} + H_2O \rightleftharpoons \text{glucose} + P_i \qquad (16)$$

Note that this reaction is quite different from the hexokinase reaction (p. 126) used for phosphorylating glucose, the equilibrium constant of which is about one thousand in favour of glucose-6-phosphate.

The glucose produced from glucose-6-phosphate leaves the liver, and is used to maintain the physiological concentration of glucose in the blood.

11 Synthesis of
ATP – the Krebs cycle

Pyruvic acid is the penultimate compound of the glycolytic pathway, and we have seen that under anaerobic conditions the $NADH_2$ produced in reaction (6) (Chapter 10) of this pathway is reoxidized to NAD by reducing pyruvic acid. In aerobic conditions, on the other hand, $NADH_2$ can be reoxidized to NAD via the respiratory chain (see p. 108) so that in this case there is no need for pyruvic acid to act as an acceptor of hydrogen. Instead, pyruvic acid, most of which (as we have seen) arises from degradation of carbohydrate, is broken down by oxidation.

The first stage in the breakdown is an oxidative decarboxylation. This reaction proceeds in several steps, catalysed by the enzyme complex pyruvate dehydrogenase. In the first step an acetyl group is removed from pyruvic acid, leaving carbon dioxide, and then this acetyl group is passed from one carrier to another until it reaches the carrier coenzyme A.

The first of these molecules that carry the acetyl group is thiamine pyrophosphate. Its structure is given on p. 27, but we shall here write it as TPP. The next carrier is lipoic acid which we shall write

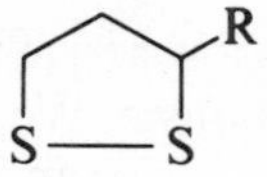

where R = $(CH_2)_4COOH$.

Coenzyme A (p. 27) we shall write CoASH. Thus as far as the formation of acetyl coenzyme A from pyruvic acid is concerned we can write the reaction sequence as follows.

$$CH_3COCOOH + TPP \rightleftharpoons CH_3CHOTPP + CO_2$$

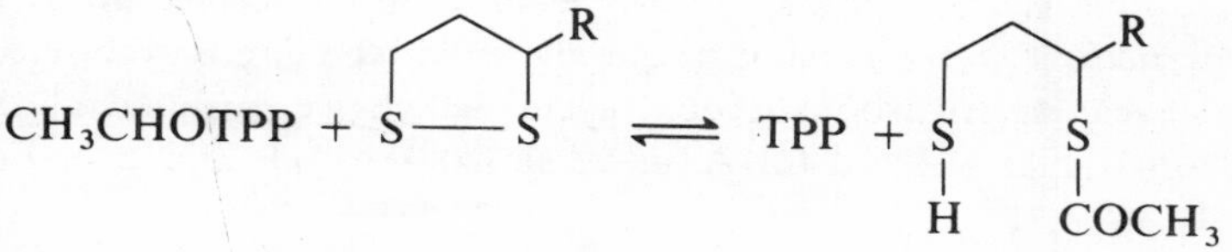

133

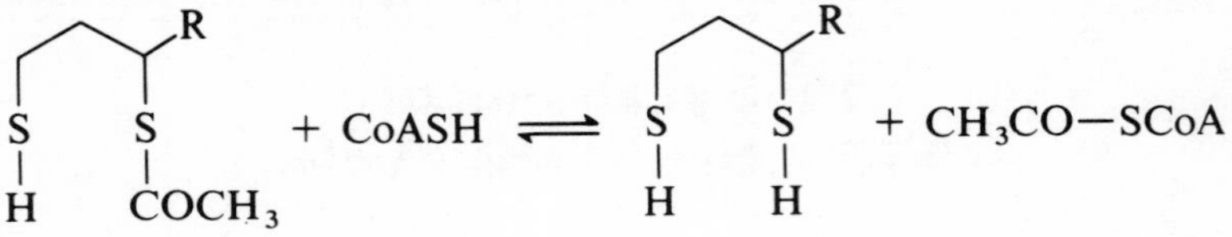

$$\text{(structure: } CH_2CH_2\text{ backbone with R, } S-H \text{ and } S-COCH_3 \text{)} + CoASH \rightleftharpoons \text{(structure with } S-H \text{ and } S-H\text{)} + CH_3CO-SCoA$$

We now have a situation similar to that which we saw in the accumulation of $NADH_2$ during glycolysis. One result of the oxidative decarboxylation of pyruvic acid is that lipoic acid becomes reduced. Since lipoic acid is a coenzyme and is present only in small quantities in the cell, it must be reoxidized if it is to function in the breakdown of further molecules of pyruvic acid. In fact reduced lipoic acid can be reoxidized by NAD in a reaction catalysed by lipoate dehydrogenase.

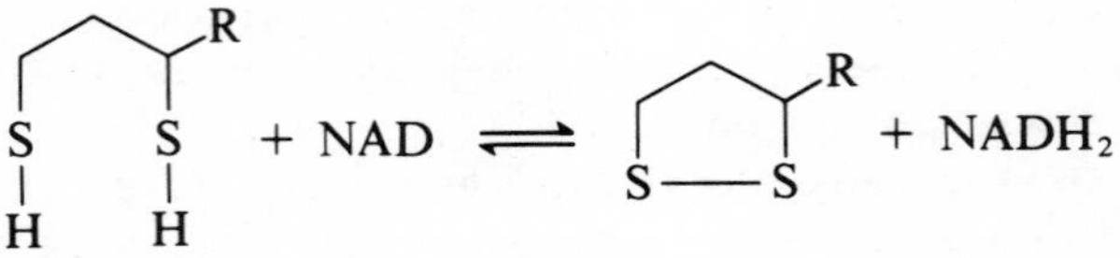

$$\text{(structure with } S-H \text{ and } S-H, \text{ R)} + NAD \rightleftharpoons \text{(cyclic structure } S-S, \text{ R)} + NADH_2$$

Since all of these reactions are proceeding in the presence of oxygen, $NADH_2$ is reoxidized by the respiratory chain. Thus the sum of the four reactions so far is

$$CH_3COCOOH + CoASH + NAD \rightleftharpoons$$

pyruvic acid

$$CH_3CO-SCoA + CO_2 + NADH_2 \quad (1)$$

acetyl coenzyme A

Acetyl coenzyme A, at which we have now arrived, is one of the central compounds of intermediary metabolism. We have just seen how it is produced by oxidation of pyruvic acid, and hence it is an intermediate in the breakdown of carbohydrates and of glycerol. We shall see later (Chapter 12) that it is the breakdown product of fatty acids, and also (Chapter 17) of some amino acids. Again, acetyl coenzyme A is the starting point for synthesis of fatty acids (see Chapter 16) and steroids and carotenoids.

We can now examine the fate of the acetyl group of acetyl coenzyme A. The oxidation of this acetyl group proceeds through a sequence of reactions that is frequently called the Krebs cycle, citric acid cycle or tricarboxylic acid cycle, which we outlined at the bottom of Figs. 9.1 and 9.2. A full version of the cycle is given in Fig. 11.1.

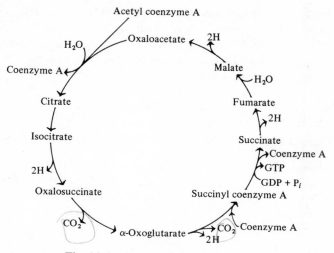

Fig. 11.1. The tricarboxylic acid cycle.

In the first reaction, acetyl coenzyme A reacts with a compound called oxaloacetic acid. This reaction is catalysed by citrate synthase.

$$CH_3CO-SCoA + \underset{\underset{CH_2COOH}{|}}{COCOOH} + H_2O \rightleftharpoons$$

oxaloacetic acid

$$\rightleftharpoons \underset{\underset{CH_2COOH}{|}}{\overset{CH_2COOH}{\underset{|}{HOCCOOH}}} + CoASH \quad (2)$$

citric acid

Oxaloacetic acid is, as we shall see shortly, formed in the *last* reaction of the Krebs cycle. Thus it acts, in a sense, as a carrier of the acetyl group through the reactions of the cycle and is left over at the end to accept another acetyl group.

The citric acid is now isomerized to form isocitric acid. The enzyme responsible for the isomerization is called aconitase.

$$\underset{\underset{CH_2COOH}{|}}{\overset{CH_2COOH}{\underset{|}{HOCCOOH}}} \rightleftharpoons \underset{\underset{CH_2COOH}{|}}{\overset{CHOHCOOH}{\underset{|}{CHCOOH}}} \quad (3)$$

citric acid   isocitric acid

Isocitric acid is dehydrogenated by NAD in a reaction catalysed by isocitrate dehydrogenase. The resulting oxalosuccinic acid is unstable and would decarboxylate spontaneously without an enzyme, but in fact the enzyme responsible for its formation also catalyses its decarboxylation.

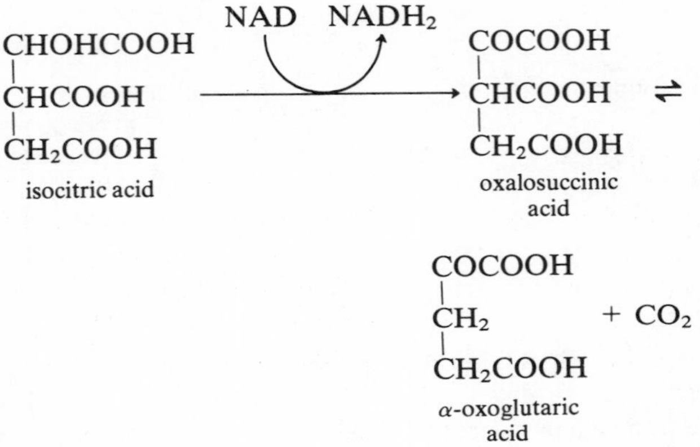

$$
\begin{array}{c}
\text{NAD} \quad \text{NADH}_2 \\
\end{array}
$$

$$
\begin{array}{ccc}
\text{CHOHCOOH} & & \text{COCOOH} \\
| & & | \\
\text{CHCOOH} & \longrightarrow & \text{CHCOOH} \quad \rightleftharpoons \\
| & & | \\
\text{CH}_2\text{COOH} & & \text{CH}_2\text{COOH} \\
\text{isocitric acid} & & \text{oxalosuccinic acid}
\end{array}
$$

$$
\begin{array}{c}
\text{COCOOH} \\
| \\
\text{CH}_2 \qquad + \text{CO}_2 \qquad (4)\\
| \\
\text{CH}_2\text{COOH} \\
\alpha\text{-oxoglutaric acid}
\end{array}
$$

The compound thus formed is α-oxoglutaric acid, which is an α-keto acid analogous to pyruvic acid. In just the same way as pyruvic acid was oxidized to acetyl coenzyme A, so α-oxoglutaric acid is oxidized to succinyl coenzyme A. The reaction involves thiamine pyrophosphate, lipoic acid and coenzyme A, and reduces NAD; the enzyme is α-oxoglutarate dehydrogenase.

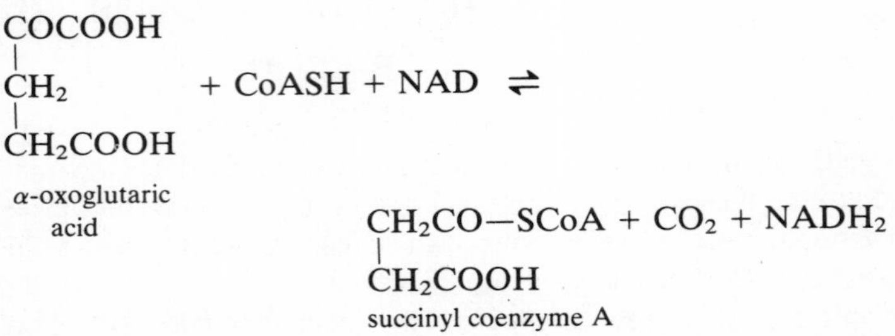

$$
\begin{array}{c}
\text{COCOOH} \\
| \\
\text{CH}_2 \qquad + \text{CoASH} + \text{NAD} \quad \rightleftharpoons \\
| \\
\text{CH}_2\text{COOH} \\
\alpha\text{-oxoglutaric acid}
\end{array}
$$

$$
\begin{array}{c}
\text{CH}_2\text{CO}-\text{SCoA} + \text{CO}_2 + \text{NADH}_2 \qquad (5)\\
| \\
\text{CH}_2\text{COOH} \\
\text{succinyl coenzyme A}
\end{array}
$$

Succinyl coenzyme A is analogous to the acetyl coenzyme A formed in reaction (1). It contains a high-energy thioester bond (see Table 7.1), and this can now be used for the synthesis of ATP – though in fact the compound first formed is GTP rather than ATP. The reaction is catalysed by a thiokinase and produces succinic acid and free coenzyme A.

$$\begin{matrix} CH_2CO-SCoA \\ | \\ CH_2COOH \end{matrix} + GDP + P_i \rightleftharpoons$$

succinyl
coenzyme A

$$\begin{matrix} CH_2COOH \\ | \\ CH_2COOH \end{matrix} + GTP + CoASH \quad (6)$$

succinic
acid

$$GTP + ADP \rightleftharpoons GDP + ATP \quad (6a)$$

Since acetyl coenzyme A also contained an energy-rich acyl mercaptide bond, it is at first sight surprising that it too was not used for the synthesis of GTP or ATP. The reason is that the energy was needed for the condensation with oxaloacetic acid to yield citric acid (reaction (2)).

The succinic acid produced in reaction (6) is oxidized by means of succinate dehydrogenase. This enzyme contains FAD as prosthetic group (see p. 108) and does not use NAD as a coenzyme.

$$\begin{matrix} CH_2COOH \\ | \\ CH_2COOH \end{matrix} + \text{enzyme-FAD} \rightleftharpoons$$

succinic
acid

$$\begin{matrix} CHCOOH \\ || \\ HCCOOH \end{matrix} + \text{enzyme-FADH}_2 \quad (7)$$

fumaric
acid

Fumaric acid accepts water in a reaction catalysed by a hydratase. The malic acid that results is oxidized by NAD with malate dehydrogenase as the enzyme.

$$\begin{matrix} CHCOOH \\ || \\ HCCOOH \end{matrix} + H_2O \rightleftharpoons \begin{matrix} CHOHCOOH \\ | \\ CH_2COOH \end{matrix} \quad (8)$$

fumaric acid malic acid

$$\begin{matrix} CHOHCOOH \\ | \\ CH_2COOH \end{matrix} + NAD \rightleftharpoons \begin{matrix} COCOOH \\ | \\ CH_2COOH \end{matrix} + NADH_2 \quad (9)$$

malic acid oxaloacetic acid

The net effect of the last three reactions is to oxidize $-CH_2-$ to $-CO-$. This is an important conversion, and we shall see (p. 141) that it occurs in the oxidation of fatty acids by an analogous sequence of three reactions.

The final reaction produces oxaloacetic acid, which can react with acetyl coenzyme A again (2). It is for this reason that the reactions just described are collectively called the Krebs *cycle*.

During the course of the reactions mentioned in this chapter, pyruvic acid is completely oxidized. The three carbon atoms of pyruvic acid are lost as carbon dioxide in reactions (1), (4) and (5). Five pairs of hydrogen atoms are released, in reactions (1), (4), (5), (7) and (9). Thus the overall reaction is formally

$$CH_3COCOOH + 3H_2O \rightarrow 3CO_2 + 5H_2.$$

Of the five pairs of hydrogen atoms, four pairs are passed to NAD. Each pair can give rise to three molecules of ATP (see p. 105). The remaining pair, produced in reaction (7), is passed to FAD and can give rise to two molecules of ATP (see p. 112). There is also a gain of one ATP (actually GTP in the first instance) in reaction (6). Thus the oxidation of each molecule of pyruvic acid can yield fifteen molecules of ATP.

We can now calculate the yield of ATP from the breakdown of glucose in aerobic conditions. We saw in the previous chapter that the glycolytic pathway produces two molecules of pyruvic acid and yields two molecules of ATP and two of $NADH_2$. In aerobic conditions these two $NADH_2$ can themselves yield six ATP because the respiratory chain will be operating. If each molecule of pyruvic acid gives fifteen ATP the total yield from the oxidation of glucose will be thirty-eight ATP.* This calculation provides a striking contrast with the yield of two ATP under anaerobic conditions.

Replenishment of Krebs cycle intermediates

The Krebs cycle is responsible for the oxidation of the acetyl group of acetyl coenzyme A, and is therefore the final common path of oxidation of carbohydrates and also (see next chapter) of fats. But the Krebs cycle intermediates have other functions too. We shall see later (p. 172) that α-oxoglutaric acid can be converted to the amino acid, glutamic acid, which itself is not only a constituent of proteins but also able to give rise to other amino acids. Oxaloacetic acid can be converted to the amino acid, aspartic acid (p. 174), and this can give rise to further amino acids and also to pyrimidines. Succinyl

---

* Some people are suspicious of this sort of calculation, since, for reasons that are not fully understood, the theoretical yield of ATP may not be achieved in all circumstances. Nonetheless, we think it useful for comparative purposes to give a rough idea of the quantity of ATP that can be obtained from substrates following different metabolic routes.

coenzyme A is involved in the synthesis of pyrroles, which are constituents of haem and of chlorophyll.

Were it not for these important synthetic side reactions, only catalytic quantities of Krebs cycle intermediates would be required to function in the oxidation of acetyl groups. But since these intermediates are constantly siphoned off for biosynthetic reactions, there must be a mechanism for replenishing the constituents of the Krebs cycle. Without such a mechanism, the terminal oxidation of carbohydrates and fats would soon fail. In part, the Krebs cycle can be replenished by the reverse of the reactions that have just been mentioned – e.g. α-oxoglutaric acid not only gives rise to glutamic acid but also can be formed from glutamic acid. But in conditions where biosynthetic reactions predominate, there must be another means of replenishing the Krebs cycle intermediates. This is provided by an ATP-dependent carboxylation of pyruvic acid catalysed by pyruvate carboxylase:

$$CH_3COCOOH + CO_2 + ATP \rightleftharpoons \underset{\substack{| \\ CH_2COOH}}{COCOOH} + ADP + P_i$$

pyruvic acid   oxaloacetic acid

The existence of this reaction has an extremely important implication. It requires that the glycolytic pathway be functioning in order to provide pyruvic acid, and thus in the absence of active carbohydrate metabolism replenishment of the Krebs cycle intermediates will be difficult or impossible. We shall return to this point in discussing the oxidation of fatty acids on p. 143.

12 Synthesis of
ATP – fat breakdown

In many organisms, a large fraction of the ATP produced comes from the oxidation of fats. As we explained in Chapter 6, fats are esters of glycerol, having the general formula

$$CH_2OCOR_1$$
$$CHOCOR_2$$
$$CH_2OCOR_3$$

and they can be readily hydrolysed by the action of lipases to produce glycerol and fatty acids.

$$
\begin{array}{ccc}
CH_2OCOR_1 & CH_2OH & R_1COOH \\
| & | & + \\
CHOCOR_2 + 3H_2O \rightleftharpoons & CHOH + R_2COOH \\
| & | & + \\
CH_2OCOR_3 & CH_2OH & R_3COOH
\end{array}
$$

We have already seen (p. 131) how glycerol can be metabolized by the reactions of the Embden–Meyerhof pathway. We shall now consider the oxidation of fatty acids.

In principle fats can contain many different fatty acids, but in practice the naturally occurring types of fatty acid are fairly few. They nearly always have an even number of carbon atoms; two of the commonest are palmitic acid ($C_{15}H_{31}COOH$) and stearic acid ($C_{17}H_{35}COOH$). We shall illustrate the oxidation of fatty acids by taking stearic acid as an example. For the present purpose it is convenient to write it as

$$CH_3(CH_2)_{12}CH_2CH_2CH_2CH_2COOH.$$

Before it can be oxidized, the fatty acid has to be activated by reaction with coenzyme A. ATP is used up, being converted to AMP rather than ADP, with the release of inorganic pyrophosphate.

$$CH_3(CH_2)_{12}CH_2CH_2CH_2CH_2COOH + CoASH + ATP \rightleftharpoons$$
$$CH_3(CH_2)_{12}CH_2CH_2CH_2CH_2CO{-}SCoA + AMP + PP_i$$

This reaction is catalysed by a thiokinase and has an equilibrium constant reasonably near to one; but the pyrophosphate produced is rapidly degraded owing to the presence of another enzyme, pyrophosphatase:

$$PP_i + H_2O \rightleftharpoons 2P_i.$$

Since the concentration of water in the cell is much higher than that of inorganic phosphate, this reaction will be driven to the right and thus there is in practice no chance of a net resynthesis of ATP from AMP and pyrophosphate. We may therefore regard the formation of fatty-acyl coenzyme A from fatty acid as effectively irreversible. We may recall that glucose too (and other sugars and glycerol) have first to be activated, by reactions that are effectively irreversible, before being further metabolized.

Stearoyl coenzyme A is now broken down by a series of reactions that are together called β oxidation. In the course of this series of reactions, the β-carbon atom of the acyl coenzyme A is gradually oxidized to yield a β-ketoacyl coenzyme A.

In the first reaction of β oxidation, a pair of <u>hydrogen atoms is removed</u> from stearoyl coenzyme A to give the corresponding α,β-unsaturated compound. The enzyme for this oxidation is fatty-acyl coenzyme A dehydrogenase; it contains FAD as a prosthetic group instead of using NAD:

$$CH_3(CH_2)_{12}CH_2CH_2CH_2CH_2CO—SCoA + enzyme\text{-}FAD \rightleftharpoons$$
$$CH_3(CH_2)_{12}CH_2CH_2CH{=}CHCO—SCoA + enzyme\text{-}FADH_2.$$

Next, the unsaturated or enoyl coenzyme A accepts water in a reaction catalysed by enoyl hydratase. This enzyme adds water specifically to form the β- (not the α-) hydroxy acyl coenzyme A.

$$CH_3(CH_2)_{12}CH_2CH_2CH{=}CHCO—SCoA + H_2O \rightleftharpoons$$
$$CH_3(CH_2)_{12}CH_2CH_2CHOHCH_2CO—SCoA.$$

This product is now oxidized again, a β-hydroxy fatty-acyl coenzyme A dehydrogenase passing a pair of hydrogen atoms to NAD.

$$CH_3(CH_2)_{12}CH_2CH_2CHOHCH_2CO—SCoA + NAD \rightleftharpoons$$
$$CH_3(CH_2)_{12}CH_2CH_2COCH_2CO—SCoA + NADH_2.$$

This series of three reactions is strikingly similar to the three reactions in the Krebs cycle by which succinic acid is oxidized to oxaloacetic acid (p. 137). The first is a dehydrogenation to produce a double bond, catalysed by an enzyme that contains FAD. The second is a hydration, and the third is an oxidation of –CHOH– to –CO– catalysed by an enzyme that uses NAD.

Now that the β-carbon atom has been completely oxidized, the resulting β-ketoacyl coenzyme A can be split with another molecule of coenzyme A. This reaction is catalysed by a β-keto thiolase.

$$CH_3(CH_2)_{12}CH_2CH_2COCH_2CO-SCoA + CoASH \rightleftharpoons$$
$$CH_3(CH_2)_{12}CH_2CH_2CO-SCoA + CH_3CO-SCoA.$$

The products are acetyl coenzyme A and the fatty-acyl coenzyme A two carbon atoms shorter than the original compound, namely palmitoyl coenzyme A. Acetyl coenzyme A enters the Krebs cycle in the normal way and palmitoyl coenzyme A re-enters the β-oxidation pathway.

$$CH_3(CH_2)_{12}CH_2CH_2CO-SCoA$$
$$\downarrow \quad \text{dehydrogenase (FAD)}$$
$$CH_3(CH_2)_{12}CH{=}CHCO-SCoA$$
$$\downarrow \quad \text{hydratase (H}_2\text{O)}$$
$$CH_3(CH_2)_{12}CHOHCH_2CO-SCoA$$
$$\downarrow \quad \text{dehydrogenase (NAD)}$$
$$CH_3(CH_2)_{12}COCH_2CO-SCoA$$
$$\downarrow \quad \text{thiolase (CoASH)}$$
$$CH_3(CH_2)_{12}CO-SCoA + CH_3CO-SCoA$$

In this way fatty acids (in the form of their acyl coenzyme A derivatives) are broken down to several acetyl coenzyme A units, each of which is then oxidized through the Krebs cycle. Stearoyl coenzyme A, for example, undergoes the β-oxidation sequence eight times and yields nine molecules of acetyl coenzyme A.

We can use these figures to calculate the yield of ATP from the complete oxidation of stearic acid. Each sequence of β oxidation involves the passage of one pair of hydrogen atoms to FAD (giving rise by oxidation to two molecules of ATP) and one pair to NAD (giving rise by oxidation to three molecules of ATP). Thus the β-oxidation sequences for stearoyl coenzyme A should yield $8 \times 5 = 40$ ATP. The oxidation of nine acetyl groups (entering the Krebs cycle as acetyl coenzyme A) should yield $9 \times 12 = 108$ ATP (see p. 138). Thus the complete oxidation of the stearoyl group yields 148 ATP molecules.

However, stearoyl coenzyme A was formed from stearic acid in a reaction in which ATP was broken down to AMP, so that the net yield would be diminished by two high-energy bonds. Notice that the first β oxidation sequence produced palmitoyl coenzyme A, not palmitic acid; so that each fatty-acyl coenzyme A derived by cleavage with β-keto thiolase can in turn undergo β oxidation: only *one* activation with ATP is needed for the entire fatty acid.

We may therefore say that the complete oxidation of stearic acid yields 146 ATP. It is instructive to compare the yield from the oxidation of stearic acid (which contains eighteen carbon atoms) with that from the oxidation of three molecules of glucose (also containing eighteen carbon atoms). Three molecules of glucose would yield $3 \times 38 = 114$ molecules of ATP. The considerably greater yield from the oxidation of stearic acid reflects the fact that a fatty acid is much more reduced than a sugar, i.e. it contains proportionately more hydrogen atoms to be oxidized.

Formation of ketone bodies

In addition to providing glucose to other tissues by hydrolysis of glucose-6-phosphate (p. 132), the liver also supplies acetoacetic acid and β-hydroxybutyric acid; these two compounds are known as ketone bodies. Acetoacetic acid is made by the condensation of two molecules of acetyl coenzyme A (which normally arise from the oxidation of fatty acids) and the deacylation of the resulting acetoacetyl coenzyme A. Reduction of acetoacetic acid gives β-hydroxybutyric acid.

$$CH_3COCH_2COOH + NADH_2 \rightleftharpoons CH_3CHOHCH_2COOH + NAD$$

acetoacetic acid $\qquad\qquad\qquad$ β-hydroxybutyric acid

These two acids travel in the blood to the other tissues, where they are reconverted to acetyl coenzyme A and metabolized through the Krebs cycle. They thus form an important source of fuel, in addition to glucose, for many tissues.

Usually the concentration of ketone bodies in the blood is fairly low, but under certain conditions the liver supplies ketone bodies to the blood at a much increased rate. This situation arises whenever carbohydrate breakdown is impaired – for example during fasting, or in diabetes. In these circumstances the animal turns to fat oxidation to supply its needs of ATP. As a result there is a large flow of acetyl coenzyme A into the Krebs cycle, which requires oxaloacetic acid for its further metabolism. Now we have mentioned previously that the cycle is an important source of biosynthetic intermediates, and that the supply of oxaloacetic acid is therefore liable to run short unless replenished. But the carboxylation reaction (p. 139) which replenishes oxaloacetic acid can only proceed if there is a supply of pyruvic acid arising from the metabolism of carbohydrates, and in circumstances when the rate of

carbohydrate metabolism is low oxaloacetic acid will be in particularly short supply.

The result is that acetyl coenzyme A, unable to enter the Krebs
cycle, will tend to be diverted in increasing quantities towards the
synthesis of ketone bodies. The presence of these compounds in the
blood in higher concentration than usual – which is known as ketosis
– is symptomatic of conditions in which the rate of carbohydrate
breakdown is diminished.

13 Synthesis of ATP and NADPH₂ – the pentose phosphate pathway

In Chapters 10 and 11 we have seen how glucose and other sugars are metabolized via the Embden–Meyerhof pathway and the Krebs cycle. These pathways account for the bulk of carbohydrate metabolism in most organisms. There is, however, an alternative pathway for the oxidation of sugars that we must now consider. It is sometimes called the Warburg–Dickens pathway, sometimes the hexose monophosphate 'shunt' and sometimes (for reasons that will become obvious) the pentose phosphate pathway.

Like glycolysis, this pathway employs glucose-6-phosphate (p. 126) as its starting material, but unlike glycolysis the pentose phosphate pathway is not a fermentation. We recall that the essential feature of a fermentation is that the molecule undergoes neither net oxidation nor net reduction; here, on the contrary, the very first reaction is an oxidation, and the pair of hydrogen atoms removed from glucose-6-phosphate is not at any stage returned to the molecule. The enzyme glucose-6-phosphate dehydrogenase passes these hydrogen atoms to NADP; we shall consider their fate later.

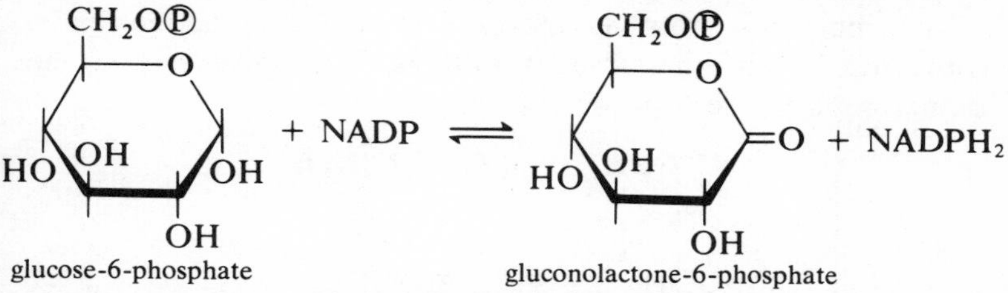

glucose-6-phosphate   gluconolactone-6-phosphate

The resulting lactone is unstable and is hydrolysed at a measurable rate in the absence of an enzyme. However, there is, in fact, a lactonase that increases the rate of reaction.

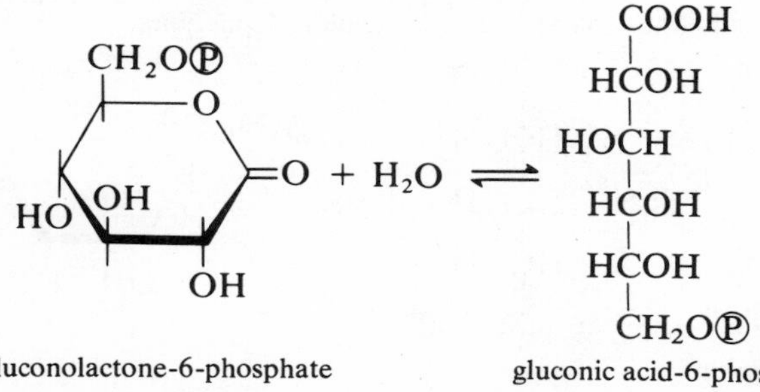

gluconolactone-6-phosphate   gluconic acid-6-phosphate

145

Next another oxidation occurs. The enzyme passes two hydrogen atoms to NADP, and the very unstable acid that results decarboxylates spontaneously. This reaction is similar to the oxidation of isocitric acid to another unstable intermediate (oxalosuccinic acid) that also decarboxylates at once (p. 136).

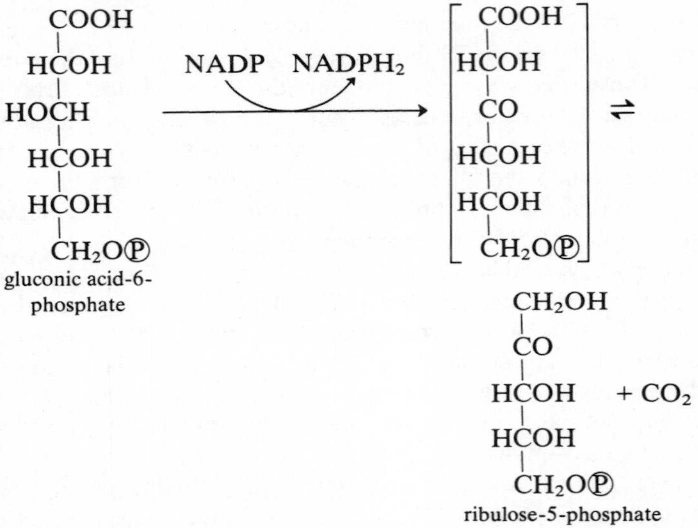

$$
\begin{array}{c}
\text{COOH} \\
\text{HCOH} \\
\text{HOCH} \\
\text{HCOH} \\
\text{HCOH} \\
\text{CH}_2\text{O}\textcircled{P}
\end{array}
\quad
\xrightarrow[\text{NADP} \quad \text{NADPH}_2]{}
\quad
\left[
\begin{array}{c}
\text{COOH} \\
\text{HCOH} \\
\text{CO} \\
\text{HCOH} \\
\text{HCOH} \\
\text{CH}_2\text{O}\textcircled{P}
\end{array}
\right]
\rightleftharpoons
$$

gluconic acid-6-phosphate

$$
\begin{array}{c}
\text{CH}_2\text{OH} \\
\text{CO} \\
\text{HCOH} \quad + \text{CO}_2 \\
\text{HCOH} \\
\text{CH}_2\text{O}\textcircled{P}
\end{array}
$$

ribulose-5-phosphate

The ribulose-5-phosphate that is formed in this reaction can be converted by an epimerase to xylulose-5-phosphate, or by an isomerase to ribose-5-phosphate.

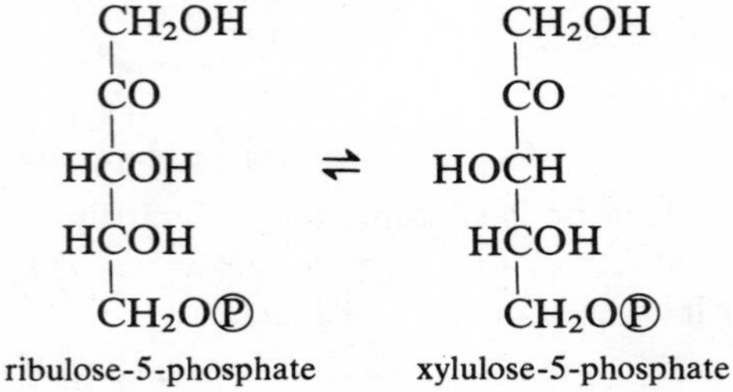

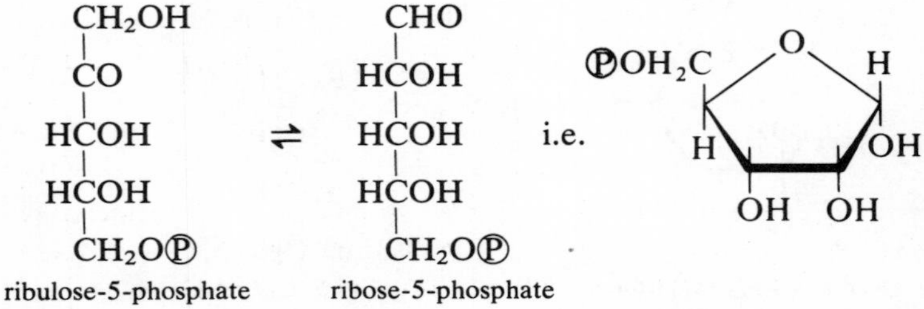

$$
\begin{array}{c}
\text{CH}_2\text{OH} \\
\text{CO} \\
\text{HCOH} \\
\text{HCOH} \\
\text{CH}_2\text{O}\textcircled{P}
\end{array}
\rightleftharpoons
\begin{array}{c}
\text{CH}_2\text{OH} \\
\text{CO} \\
\text{HOCH} \\
\text{HCOH} \\
\text{CH}_2\text{O}\textcircled{P}
\end{array}
$$

ribulose-5-phosphate   xylulose-5-phosphate

$$
\begin{array}{c}
\text{CH}_2\text{OH} \\
\text{CO} \\
\text{HCOH} \\
\text{HCOH} \\
\text{CH}_2\text{O}\textcircled{P}
\end{array}
\rightleftharpoons
\begin{array}{c}
\text{CHO} \\
\text{HCOH} \\
\text{HCOH} \\
\text{HCOH} \\
\text{CH}_2\text{O}\textcircled{P}
\end{array}
\quad \text{i.e.}
$$

ribulose-5-phosphate   ribose-5-phosphate

Before discussing the further metabolism of these pentose phosphates, we may stop to consider what are the particular functions of the reactions we have described so far. They appear to represent an alternative means for the breakdown of glucose, different from the Embden–Meyerhof–Krebs pathways. But why should a second route be needed for glucose breakdown? We have already seen that the Embden–Meyerhof pathway permits the degradation of glucose in anaerobic conditions, and that the same pathway combined with the Krebs cycle results in the total oxidation of glucose to carbon dioxide and water. As far as the formation of ATP is concerned, these routes provide adequate means of breaking down glucose; so it seems surprising that another pathway should exist alongside the Embden–Meyerhof–Krebs route.

To find out the significance of the reactions leading from glucose-6-phosphate to the pentose phosphates, let us look more closely at the compounds they produce.

In the first place, these reactions lead to pentoses and in particular to ribose. Ribose is a constituent of such coenzymes as NAD and ATP, and, of course, of the nucleotides in RNA. From the nucleoside triphosphates (ATP, CTP, etc.) that are used in the synthesis of RNA, deoxynucleoside triphosphates (deoxyATP, deoxyCTP, etc.) can be formed; these are the building blocks for the synthesis of DNA (deoxyribonucleic acid). Some of these conversions will be described in Chapter 19.

Secondly, the oxidation of glucose-6-phosphate yields not only pentose phosphates but also $NADPH_2$. We saw in Chapter 8 that $NADH_2$ is for the most part reoxidized by the respiratory chain and that this oxidation yields ATP. On the other hand, $NADPH_2$ is used primarily as a reducing agent (see Chapter 7). Just as the complete breakdown of large molecules (e.g. lipids and polysaccharides) involves their oxidation, so the synthesis of these molecules from simple precursors requires reduction. Fatty acids, for example, are synthesized starting from acetyl coenzyme A (Chapter 16); since the acetyl group is CH_3CO-, whereas fatty acids have long chains of the type $CH_3CH_2CH_2CH_2...$, it is evident that reduction as well as condensation is required in the synthesis: we shall see in detail (Chapter 16) how this reduction by $NADPH_2$ takes place. $NADPH_2$ is also used in the synthesis of steroids and in other important reductions.

A tissue that is actively synthesizing fats or steroids will require large quantities of $NADPH_2$. We have seen that the oxidation of glucose-6-phosphate to ribulose-5-phosphate reduces two molecules of NADP to $NADPH_2$, but two molecules of $NADPH_2$ suffice only for the reduction of one acetyl group to $-CH_2CH_2-$ in a fatty acid (see p. 168). To synthesize (for example) stearic acid requires sixteen molecules of $NADPH_2$, i.e. the oxidation of eight molecules of glucose-6-phosphate to eight molecules of ribulose-5-phosphate.

We have just pointed out that a supply of pentoses is extremely important; but there can be too much of a good thing, and for a tissue rapidly synthesizing fats or steroids the quantity of ribulose-5-phosphate produced as a necessary by-product of reducing NADP may well be an embarrassment. There is, however, a means of converting pentose back to hexose by shuffling around the carbon atoms of various sugars and thus forming five hexoses from six pentoses. These reactions for the interconversion of sugars form the second part of the Warburg–Dickens pathway, and we shall see (Chapter 15) that similar reactions are used in the interconversion of sugars after the fixation of carbon dioxide in photosynthesis.

Two special enzymes are involved in these interconversions. Transketolase catalyses the transfer of the CH_2OHCO- group from a ketose to an aldose, and transaldolase the transfer of the $CH_2OHCOCHOH-$ group, also from a ketose to an aldose. Thus the transketolase transfers a two-carbon fragment and the transaldolase a three-carbon fragment, and between them they can accomplish the conversion of three molecules of pentose phosphate into two molecules of hexose phosphate and one of triose phosphate:

$$\mathbf{C_5} + \mathbf{C_5} \xrightleftharpoons{\text{transketolase}} C_7 + C_3$$

$$C_7 + C_3 \xrightleftharpoons{\text{transaldolase}} C_4 + C_6$$

$$\mathbf{C_5} + C_4 \xrightleftharpoons{\text{transketolase}} C_3 + C_6$$

(Bold type signifies the original pentose phosphate molecules.)
    Writing the reactions in more detail gives:

xylulose-5-Ⓟ + ribose-5-Ⓟ ⇌

sedoheptulose-7-Ⓟ + glyceraldehyde-3-phosphate

sedoheptulose-7-Ⓟ + glyceraldehyde-3-phosphate ⇌

erythrose-4-Ⓟ + fructose-6-Ⓟ

xylulose-5-Ⓟ + erythrose-4-Ⓟ ⇌

glyceraldehyde-3-phosphate + fructose-6-Ⓟ

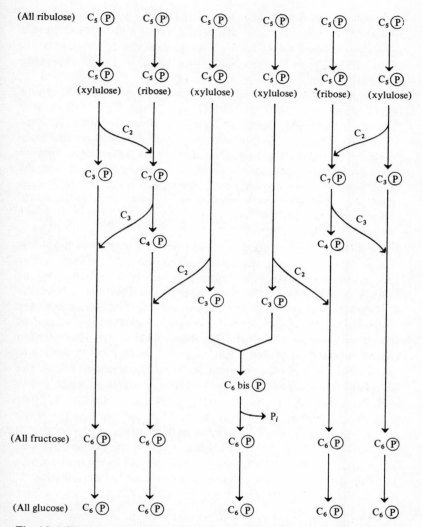

Fig. 13.1. The interconversion of sugars in the pentose phosphate pathway.

If the whole of this sequence is carried out twice, the six molecules of pentose phosphate are converted into four molecules of fructose-6-phosphate and two molecules of glyceraldehyde-3-phosphate. It is formally possible for the two molecules of glyceraldehyde-3-phosphate to give rise to fructose-1,6-bisphosphate, by the reverse of reactions (4) and (5) of the Embden–Meyerhof pathway (p. 127);

fructose-1,6-bisphosphate can then be cleaved by its specific phosphatase (p. 158) to give fructose-6-phosphate. Thus, as we mentioned above, the sequence provides for the conversion of six molecules of pentose phosphate into five molecules of hexose phosphate.

We have seen (p. 146) that the original pentose phosphate that derived from the oxidation of glucose-6-phosphate was ribulose-5-phosphate, which then became converted into xylulose-5-phosphate and ribose-5-phosphate. In a similar way, the fructose-6-phosphate which is the product of the rearrangements by transketolase and transaldolase can be converted into glucose-6-phosphate by hexose phosphate isomerase (p. 126). It is thus possible to summarize the reactions as

$$6 \times \text{ribulose-5-phosphate} \rightleftharpoons 5 \times \text{glucose-6-phosphate} + P_i$$

and a flow-sheet illustrating these conversions is given in Fig. 13.1.

We should emphasize that, although all these reactions can occur and, in some circumstances, certainly do, the metabolic conversions are not nearly as formal as they appear here. Frequently both the Embden–Meyerhof and the Warburg–Dickens pathways are operating in the same cell at the same time. Thus the fructose-6-phosphate formed by the transketolase and transaldolase reaction from ribulose-5-phosphate may well not be converted back into glucose-6-phosphate but instead undergo glycolysis. Similarly the glyceraldehyde-3-phosphate derived from pentose phosphate may well be oxidized to glyceric acid-1,3-bisphosphate (p. 128) or perhaps give rise to glycerol (p. 131). The chief importance of the first part of the Warburg–Dickens pathway appears to be that it provides pentose and $NADPH_2$; the importance of the reactions summarized in the flow-sheet is that they provide a means by which surplus pentose can, if necessary, be converted back to hexose.

14 Synthesis of ATP and NADPH₂ – the light reaction of photosynthesis

In the last four chapters we have seen how heterotrophic organisms (those that live by breaking down organic compounds) derive the bulk of their ATP and $NADPH_2$. But heterotrophic organisms depend ultimately on autotrophic organisms, which are capable of synthesizing organic compounds from simple inorganic materials. It is on the ability of autotrophic organisms to fix carbon dioxide into sugars that all life on the earth depends. In this chapter and the next we shall describe in outline how this fixation occurs.

This chapter is devoted to studying how the prerequisites of carbon dioxide fixation are provided. If we consider the conversion

$$CO_2 \rightarrow (CH_2O)_n$$

where $(CH_2O)_n$ represents a carbohydrate (e.g. glucose), it is clear that there are two prerequisites. The reaction involves the building up of a very small molecule into a larger molecule, and therefore requires ATP. It also involves reduction, and therefore requires a reducing agent (which turns out to be $NADPH_2$). Now the fixation of carbon dioxide into carbohydrates is a process largely confined to green plants and certain pigmented micro-organisms, and it is dependent on the energy of sunlight. So the problem that we have to consider in this chapter is, how can light energy be used to provide ATP and to reduce NADP?

The processes by which ATP is synthesized and NADP reduced are often known as the 'light reaction' of photosynthesis, and they are absolutely dependent on the presence of characteristic photosynthetic pigments. These pigments are confined in plants to the structures known as chloroplasts (see p. 99), which can carry out the light reaction even after isolation from the plant. The characteristic pigments of the chloroplasts are various chlorophylls (usually identified by their absorption spectra) and carotenoids, and in addition there are special cytochromes that occur only in plants.

The majority of the chlorophyll molecules in plants belong to the molecular species chlorophyll a, and these tend to be organized into arrays of several hundred molecules. Many of these arrays are associated with one molecule of a specialized chlorophyll molecule called P700 and with a molecule of the specialized plant cytochrome called cytochrome *f*. A complex of this sort forms one unit of a system called 'pigment system I', which is believed to be involved in the primary photochemical event by which ATP is produced.

Although the details of the light reaction in photosynthesis are by no means agreed, there is evidence to support some such scheme as that set out in Fig. 14.1. Light falling on any of the chlorophyll *a*

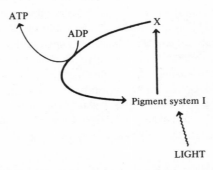

Fig. 14.1. The synthesis of ATP in the chloroplast. Bold arrows represent flow of electrons.

molecules in pigment system I causes an electron from the associated chlorophyll P700 to be promoted to a higher energy level. This event causes P700 to become an extremely good reducing agent, capable now of donating an electron to an acceptor (called X because it is unidentified), which normally P700 would be unable to reduce. From the reduced X thus formed, the electron can be transferred to an electron transport chain. This chain consists of cytochromes and other electron carriers, and it is probably quite similar to the electron transport chain of oxidative phosphorylation. At all events, during the passage of an electron along the chain ATP is synthesized from ADP. The final acceptor of the electron, however, is not oxygen but the pigment system I itself, which had been left electron deficient by the light-induced event. Thus in this system ATP is produced by a cyclical flow of electrons, and the process is sometimes called 'cyclic photophosphorylation'.

This description can account for ATP synthesis, but it does not explain the formation of the other prerequisite of carbon dioxide fixation, $NADPH_2$. For the reduction of NADP (as well as for further synthesis of ATP) a rather more complicated series of

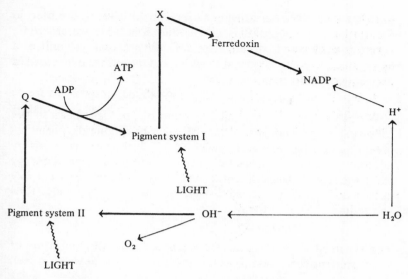

Fig. 14.2. The synthesis of ATP and the reduction of NADP in the chloroplast.

reactions appears to take place: this is illustrated in Fig. 14.2. Once again an electron in pigment system I is promoted to a higher energy level by light and passed to X; but the reduced X, instead of donating an electron to the electron transport chain, reduces the iron-containing protein ferredoxin which in turn passes an electron to NADP. However, pigment system I is now deficient in an electron, and in order for it to be reduced again a second photochemical event is necessary. This takes place in 'pigment system II', which is a complex containing the specialized chlorophyll called P670. In response to light this chlorophyll loses an electron to an acceptor known as Q, which in turn transfers an electron to pigment system I. Just as before, the electron is passed along an electron transport chain, and this passage is accompanied by the synthesis of ATP.

As a result of these events, NADP has acquired an electron and pigment system II has lost one; the intermediate carriers have returned to their previous state. Now in order to form $NADPH_2$ from NADP, not only electrons but also H^+ ions are required; moreover, in order to restore pigment system II to its normal state another electron is required. Both of these requirements are met from the products of ionization of water. H^+ ions from water, together with electrons arising via ferredoxin from X, are used in the formation of $NADPH_2$. The other ion from water, OH^-, donates an electron to the electron-deficient pigment system II, and as a result becomes converted to water and molecular oxygen. This may be written formally as

$$4OH^- \rightarrow 2H_2O + O_2 + 4e^-.$$

We can summarize the light reaction of photosynthesis in the following way. The absorption of light by pigments promotes electrons to a higher energy level, at which they are received by acceptors. As the electrons fall back again to their original energy level, they can do work in synthesizing ATP, in the same sort of way as they do in oxidative phosphorylation. Alternatively the electrons can reduce NADP in the reaction

$$NADP + 2e + 2H^+ \rightarrow NADPH_2.$$

The H^+ needed for this reaction is derived from the ionization of water, and the corresponding OH^- loses an electron as we have seen above.

This description is simplified, and (because our understanding of the light reaction is still very incomplete) it may possibly be incorrect. It does, however, account for the characteristic features of photosynthesis in green plants: the synthesis of ATP and the reduction of NADP, and the concomitant production of molecular oxygen.

15 Uses of ATP and NADPH₂ – the synthesis of polysaccharides

Armed with ATP and NADPH₂, the synthesis of which we discussed in the last chapter, plants are capable of fixing carbon dioxide into sugars and thus, ultimately, of synthesizing starch. In this chapter we shall describe some of the details of these processes, which are often called the 'dark reactions' of photosynthesis.

We can write the outline of carbon dioxide fixation in this way.

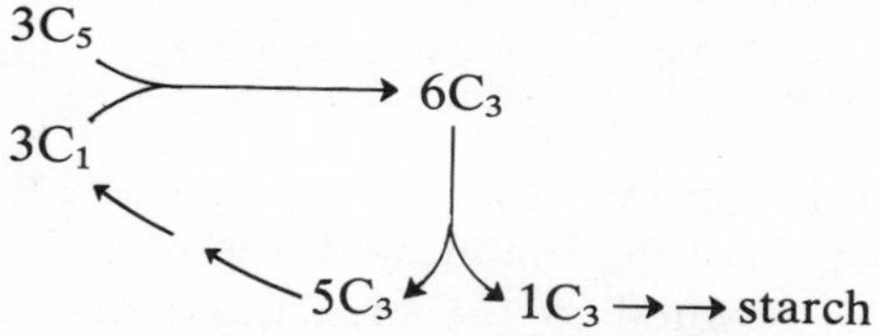

$$3C_5 \searrow$$
$$3C_1 \nearrow \longrightarrow 6C_3$$
$$\nwarrow \swarrow \searrow$$
$$5C_3 \quad 1C_3 \rightarrow \rightarrow \text{starch}$$

The key reaction is that in which carbon dioxide reacts with a pentose to yield two molecules of triose. If we consider this reaction as multiplied by three, the yield is six molecules of triose. Of these six, one molecule is built up to glucose and eventually to starch; the remaining five are rearranged to form three molecules of the pentose which can again accept carbon dioxide. These processes are sometimes called the Calvin cycle.

The Calvin cycle

The pentose with which carbon dioxide reacts is the bisphosphate derivative of ribulose. The enzyme is carboxydismutase, and the reaction yields two molecules of glyceric acid-3-phosphate.

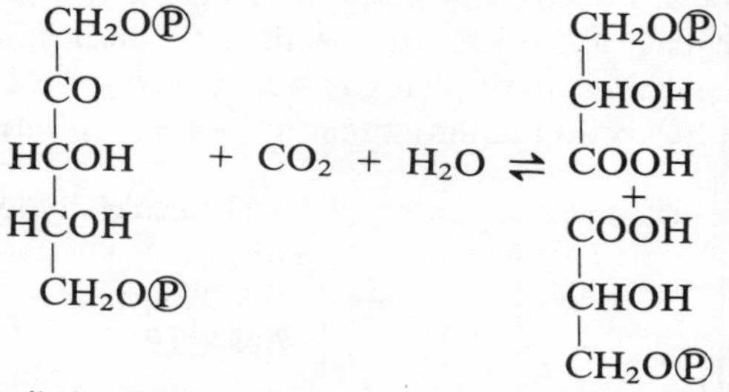

$$
\begin{array}{l}
\text{CH}_2\text{O}\circledP \\
| \\
\text{CO} \\
| \\
\text{HCOH} \quad + \text{CO}_2 + \text{H}_2\text{O} \rightleftharpoons \\
| \\
\text{HCOH} \\
| \\
\text{CH}_2\text{O}\circledP
\end{array}
\qquad
\begin{array}{l}
\text{CH}_2\text{O}\circledP \\
| \\
\text{CHOH} \\
| \\
\text{COOH} \\
+ \\
\text{COOH} \\
| \\
\text{CHOH} \\
| \\
\text{CH}_2\text{O}\circledP
\end{array}
\qquad (1)
$$

ribulose-1,5-bisphosphate   2 × glyceric acid-3-phosphate

The glyceric acid-3-phosphate is now phosphorylated by ATP in the presence of a kinase, and the resulting glyceric acid-1,3-bisphosphate is reduced by $NADPH_2$ to glyceraldehyde-3-phosphate; the enzyme is triose phosphate dehydrogenase.

$$
\begin{array}{ll}
\text{COOH} & \text{COO\textcircled{P}} \\
| & | \\
\text{CHOH} + \text{ATP} \rightleftharpoons \text{CHOH} + \text{ADP} \\
| & | \\
\text{CH}_2\text{O\textcircled{P}} & \text{CH}_2\text{O\textcircled{P}}
\end{array} \qquad (2)
$$

glyceric acid-3-phosphate   glyceric acid-1,3-bisphosphate

$$
\begin{array}{ll}
\text{COO\textcircled{P}} & \text{CHO} \\
| & | \\
\text{CHOH} + \text{NADPH}_2 \rightleftharpoons \text{CHOH} + \text{NADP} + \text{P}_i \\
| & | \\
\text{CH}_2\text{O\textcircled{P}} & \text{CH}_2\text{O\textcircled{P}}
\end{array} \qquad (3)
$$

glyceric acid-1,3-bisphosphate   glyceraldehyde-3-phosphate

In essence, these last two reactions are precisely the reverse of reactions (6) and (7) of the Embden–Meyerhof pathway (p. 128). That pathway is devoted to the breakdown of carbohydrates; photosynthesis is devoted to the formation of carbohydrates. Reaction (6) of the Embden–Meyerhof pathway is an oxidation; reaction (3) of the Calvin cycle is a corresponding reduction (note, however, that it uses $NADPH_2$ (pp. 104 and 105)). Reaction (7) of the Embden–Meyerhof pathway synthesizes ATP; reaction (2) of the Calvin cycle uses ATP.

In reactions (2) and (3) of the Calvin cycle we can therefore see, in part, how ATP and $NADPH_2$, formed in the light reaction, are used in the dark reaction. We shall discover soon (p. 158) that there is a further requirement for ATP in the rearrangement leading to ribulose-1,5-bisphosphate.

Before considering the details of this rearrangement, we ought to have a look at the stoichiometry of reactions (1), (2) and (3). It is convenient (see above) to start with three molecules of ribulose bisphosphate; thus, multiplying reaction (1) by three and reactions (2) and (3) by six and adding them up, we get equation (4).

$$
\begin{array}{ll}
3 \times \text{ribulose-1,5-bisphosphate} & 6 \times \text{glyceraldehyde-3-phosphate} \\
+3\text{CO}_2 & +6\text{P}_i \\
+3\text{H}_2\text{O} \rightarrow & +6\text{ADP} \\
+6\text{ATP} & +6\text{NADP} \\
+6\text{NADPH}_2 &
\end{array} \qquad (4)
$$

Of these six molecules of glyceraldehyde-3-phosphate, five represent merely a rearrangement of three molecules of ribulose bisphosphate, but one represents a net gain to the system. In other words one molecule of triose phosphate on the right-hand side of this sum has been synthesized from the three molecules of carbon dioxide on the left-hand side. It is the achievement of this synthesis that is characteristic of photosynthesis. By contrast the reactions involved in building up sugars and polysaccharides from triose phosphate are essentially similar in photosynthetic and non-photosynthetic organisms; we shall follow these reactions later.

The remaining five molecules of triose phosphate can now be rearranged. Many of the reactions involved are similar to those we described in the Warburg–Dickens pathway (Chapter 13) – the transketolase (p. 148) used in that pathway is required here as well (although the transaldolase is not). In addition triose phosphate isomerase is needed to interconvert glyceraldehyde-3-phosphate and dihydroxyacetone phosphate.

$$
\begin{array}{ccc}
\text{CHO} & & \text{CH}_2\text{OH} \\
| & & | \\
\text{CHOH} & \rightleftharpoons & \text{CO} \qquad\qquad (5) \\
| & & | \\
\text{CH}_2\text{O}\textcircled{P} & & \text{CH}_2\text{O}\textcircled{P}
\end{array}
$$

glyceraldehyde-3-   dihydroxyacetone
phosphate   phosphate

We also need the enzyme aldolase. We saw previously (p. 126) that this enzyme can catalyse the breakdown of fructose-1,6-bisphosphate to two molecules of triose phosphate. The same enzyme can, in fact, catalyse the reverse reaction, and, more generally, it can carry out the aldol condensation between dihydroxyacetone phosphate and various aldehydes.

In outline, the conversions are as follows:

$$\mathbf{C_3} + \mathbf{C_3} \xrightleftharpoons{\text{aldolase}} C_6, \qquad\qquad (6)$$

$$C_6 + \mathbf{C_3} \xrightleftharpoons{\text{transketolase}} C_4 + C_5, \qquad\qquad (7)$$

$$C_4 + \mathbf{C_3} \xrightleftharpoons{\text{aldolase}} C_7, \qquad\qquad (8)$$

$$C_7 + \mathbf{C_3} \xrightleftharpoons{\text{transketolase}} C_5 + C_5. \qquad\qquad (9)$$

(Bold type signifies the original triose phosphate molecules.)

Writing the reactions in more detail gives:

glyceraldehyde-3-phosphate + dihydroxyacetone phosphate $\rightleftharpoons$

fructose-1,6-bisphosphate   (6)

fructose-1,6-bisphosphate + H_2O $\rightleftharpoons$

fructose-6-phosphate + P_i (6a)

fructose-6-phosphate + glyceraldehyde-3-phosphate $\rightleftharpoons$

erythrose-4-phosphate + xylulose-5-phosphate   (7)

erythrose-4-phosphate + dihydroxyacetone phosphate $\rightleftharpoons$

sedoheptulose-1,7-bisphosphate   (8)

sedoheptulose-1,7-bisphosphate + H_2O $\rightleftharpoons$

sedoheptulose-7-phosphate + P_i (8a)

sedoheptulose-7-phosphate + glyceraldehyde-3-phosphate $\rightleftharpoons$

ribose-5-phosphate + xylulose-5-phosphate   (9)

Each of the two xylulose-5-phosphate molecules formed (reactions (7) and (9)) can be converted to ribulose-5-phosphate. Similarly, the ribose-5-phosphate produced in reaction (9) can also be converted to ribulose-5-phosphate. The enzymes responsible for these conversions have been mentioned on p. 146. A complete flow-sheet for the interconversions is given in Fig. 15.1.

We may note that two of the above reactions, (6a) and (8a), involve the hydrolysis of a sugar bisphosphate by a specific phosphatase to yield a sugar monophosphate. In consequence, the conversions shown in reactions (6) to (9) give rise to only three molecules of organic phosphate from five molecules of organic phosphate; the remaining two phosphate groups are lost by hydrolysis. In other words, the three molecules of ribulose produced are in the form of ribulose monophosphate. However, the fixation of carbon dioxide (reaction (1)) requires ribulose bisphosphate. Each ribulose-5-phosphate must therefore be converted to ribulose-1,5-bisphosphate at the expense of ATP, as we discussed on p. 102. The reaction is catalysed by a kinase:

ribulose-5-phosphate + ATP $\rightleftharpoons$

ribulose-1,5-bisphosphate + ADP.   (10)

Now that we have arrived at ribulose-1,5-bisphosphate, which can react again with carbon dioxide, we have completed an account of the reactions involved in the synthesis of triose phosphate. We can now look once more at the stoichiometry of the Calvin cycle.

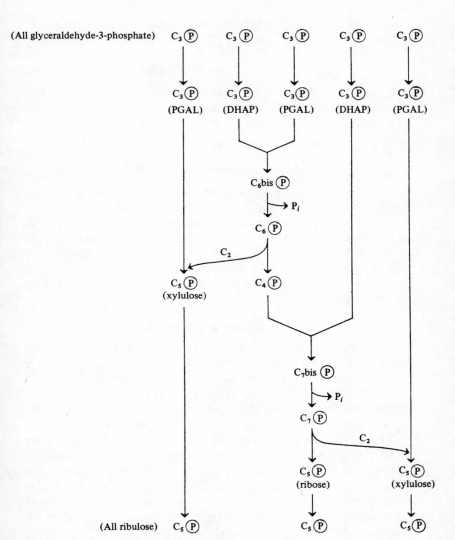

Fig. 15.1. The interconversions of sugars in the Calvin cycle. PGAL = glyceraldehyde-3-phosphate. DHAP = dihydroxyacetone phosphate.

Summing reaction (6), (6a), (7), (8), (8a) and (9) gives:

$$5 \times \text{triose phosphate} + 2H_2O \rightleftharpoons 3 \times \text{pentose phosphate} + 2P_i$$

$$(11)$$

and multiplying reaction (10) by three gives:

$3 \times$ pentose phosphate $+ 3$ATP $\rightarrow$

$$3 \times \text{pentose bisphosphate} + 3\text{ADP}. \qquad (12)$$

The sum of reactions (4), (11) and (12) gives the complete reaction for the fixation of three molecules of carbon dioxide to form one molecule of triose phosphate.

$3CO_2 + 5H_2O + 9ATP + 6NADPH_2 \rightarrow$

$$\text{glyceraldehyde-3-phosphate} + 9\text{ADP} + 8P_i + 6\text{NADP}.$$

This summary reaction represents a simple biosynthesis – the first we have so far considered in this book. Three points emerge very clearly from it. In the first place, it uses ATP to build a very simple molecule up into a larger molecule. Secondly, it uses $NADPH_2$ to reduce a highly oxidized molecule. Thirdly, it includes a number of reactions ((6), (8) and (10)) which can be regarded as 'irreversible', so that the equilibrium of the process as a whole is strongly in favour of the biosynthetic product.

The biosynthesis of polysaccharides

In the context of photosynthesis, the synthesis of polysaccharide means the use of triose phosphate to build up starch. But it is convenient at this stage to broaden the context. Polysaccharides are synthesized by almost all organisms, whether photosynthetic or not. In animals, for example, glycogen is synthesized during rest from muscular activity from the lactic acid that has accumulated (p. 130). Glycogen can be synthesized too from some amino acids – those that give rise to pyruvic acid, such as alanine or cysteine, and those that give rise to Krebs cycle intermediates, such as aspartic acid and glutamic acid (see Chapter 17). The synthesis of polysaccharide from these precursors proceeds *via* triose phosphate (see reactions (20) and (21) below); and the triose phosphate produced in the Calvin cycle joins the pathway at this stage.

We can for convenience regard the synthesis of polysaccharide as taking place in two stages – the synthesis of glucose-1-phosphate from smaller molecules, and then the polymerization of glucose residues into a polysaccharide chain. The synthesis of glucose-1-phosphate makes use of reactions most of which we have in fact previously mentioned, but it is useful to bring them together here. The reactions of polymerization are new ones, which we shall discuss in greater detail.

If we start from lactic acid, the first reaction is an oxidation by NAD with lactate dehydrogenase.

$$
\begin{array}{cc}
CH_3 & CH_3 \\
| & | \\
CHOH + NAD \rightleftharpoons CO & + NADH_2 \\
| & | \\
COOH & COOH \\
\text{lactic acid} & \text{pyruvic acid}
\end{array}
\tag{13}
$$

This is simply the reverse of the reaction given on p. 130; and, just as in the glycolysis pathway the $NADH_2$ needed to reduce pyruvic acid is derived from an oxidation earlier in the pathway, so here the $NADH_2$ produced will be used later in the pathway at reaction (19).

In glycolysis, pyruvic acid is formed from phospho-*enol*pyruvic acid (p. 129) but because of the high negative $\Delta G^{0\prime}$ of that reaction a bypass is needed in order to form phospho-*enol*pyruvic acid from pyruvic acid. This is provided by carboxylation of pyruvic acid to oxaloacetic acid followed by decarboxylation to phospho-*enol*-pyruvic acid. The carboxylation reaction was described on p. 139; the decarboxylation is catalysed by an enzyme called phospho-*enol*pyruvate carboxykinase and involves the splitting of GTP.

$$
\begin{array}{c}
CH_3COCOOH + CO_2 + ATP \rightleftharpoons COCOOH + ADP + P_i \\
| \\
CH_2COOH
\end{array}
\tag{14}
$$

$$
\text{pyruvic acid} \qquad\qquad \text{oxaloacetic acid}
$$

$$
\begin{array}{cc}
COCOOH & CH_2 \\
| & || \\
CH_2COOH + GTP \rightleftharpoons CO\textcircled{P} & + GDP + CO_2 \\
& | \\
& COOH
\end{array}
\tag{15}
$$

$$
\text{oxaloacetic acid} \qquad \text{phospho-}enol\text{pyruvic}
$$
$$
\text{acid}
$$

Reaction (15) also provides a means by which the carbon atoms of those amino acids that give rise to Krebs cycle intermediates (see Chapter 17) can enter the synthetic sequence.

Phospho-*enol*pyruvic acid accepts water in a reaction catalysed by enolase, and the resulting glyceric acid-2-phosphate is converted by a mutase to glyceric acid-3-phosphate.

$$
\begin{array}{cc}
CH_2 & CH_2OH \\
|| & | \\
CO\textcircled{P} + H_2O \rightleftharpoons CHO\textcircled{P} & \\
| & | \\
COOH & COOH
\end{array}
\tag{16}
$$

$$
\text{phospho-}enol\text{pyruvic} \qquad \text{glyceric acid-2-}
$$
$$
\text{acid} \qquad\qquad \text{phosphate}
$$

$$
\begin{array}{ccc}
\mathrm{CH_2OH} & & \mathrm{CH_2O\textcircled{P}} \\
| & & | \\
\mathrm{CHO\textcircled{P}} & \rightleftharpoons & \mathrm{CHOH} \\
| & & | \\
\mathrm{COOH} & & \mathrm{COOH}
\end{array} \qquad (17)
$$

glyceric acid-2-   glyceric acid-3-
phosphate   phosphate

Glyceric acid-3-phosphate is now phosphorylated by ATP in the presence of a kinase, and the glyceric acid-1,3-bisphosphate is reduced by $NADH_2$ in the presence of triose phosphate dehydrogenase.

$$
\begin{array}{ccc}
\mathrm{CH_2O\textcircled{P}} & & \mathrm{CH_2O\textcircled{P}} \\
| & & | \\
\mathrm{CHOH} + \mathrm{ATP} & \rightleftharpoons & \mathrm{CHOH} + \mathrm{ADP} \\
| & & | \\
\mathrm{COOH} & & \mathrm{COO\textcircled{P}}
\end{array} \qquad (18)
$$

glyceric acid-3-   glyceric acid-1,3-
phosphate   bisphosphate

$$
\begin{array}{ccc}
\mathrm{CH_2O\textcircled{P}} & & \mathrm{CH_2O\textcircled{P}} \\
| & & | \\
\mathrm{CHOH} + \mathrm{NADH_2} & \rightleftharpoons & \mathrm{CHOH} + \mathrm{NAD} + P_i \\
| & & | \\
\mathrm{COO\textcircled{P}} & & \mathrm{CHO}
\end{array} \qquad (19)
$$

glyceric acid-1,3-   glyceraldehyde-3-
bisphosphate   phosphate

These last four reactions are simply the reversal of the four corresponding reactions in glycolysis (pp. 128–9).

If the starting point for the synthesis of polysaccharide is lactic acid, the $NADH_2$ needed in reaction (19) is generated by the oxidation reaction (13). If, on the other hand, the starting point for the synthesis is pyruvic acid or oxaloacetic acid, or one of their amino-acid precursors, then the $NADH_2$ must be provided from elsewhere. The synthesis of polysaccharide from one of these latter compounds is another example of the need for reduction in many biosynthetic reactions. Reactions (18) and (19) are parallel with reactions (2) and (3) of the Calvin cycle (p. 156), the only difference being that the reduction during photosynthesis uses $NADPH_2$ instead of $NADH_2$.

Glyceraldehyde-3-phosphate produced in reaction (19) is interconverted with dihydroxyacetone phosphate, the interconversion being catalysed by triose phosphate isomerase. The two triose phosphates can now react to form fructose-1,6-bisphosphate in the presence of aldolase.

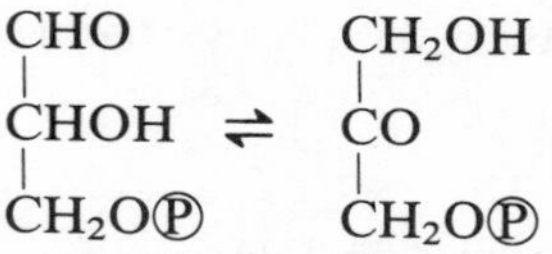

$$
\begin{array}{ccc}
\text{CHO} & & \text{CH}_2\text{OH} \\
| & & | \\
\text{CHOH} & \rightleftharpoons & \text{CO} \\
| & & | \\
\text{CH}_2\text{O}\textcircled{P} & & \text{CH}_2\text{O}\textcircled{P}
\end{array} \qquad (20)
$$

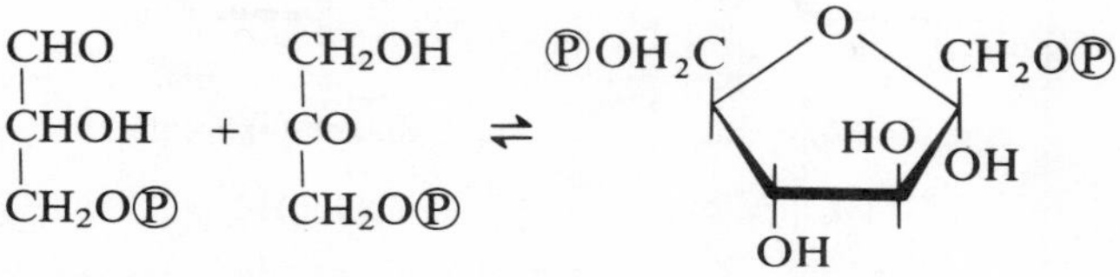

glyceraldehyde-3-   dihydroxyacetone
phosphate   phosphate

$$
\begin{array}{ccc}
\text{CHO} & \text{CH}_2\text{OH} \\
| & | \\
\text{CHOH} & + \quad \text{CO} & \rightleftharpoons \\
| & | \\
\text{CH}_2\text{O}\textcircled{P} & \text{CH}_2\text{O}\textcircled{P}
\end{array}
$$

$$(21)$$

glyceraldehyde-   dihydroxyacetone   fructose-1,6-
3-phosphate   phosphate   bisphosphate

These two reactions are again reversals of the corresponding reactions ((5) and (4)) of glycolysis (p. 127). They also correspond to reactions (5) and (6) of the Calvin cycle (p. 157). Thus the triose phosphate that was produced by photosynthesis can be regarded as entering the sequence at this point.

Fructose-1,6-bisphosphate is now cleaved in another reaction that also occurs in the Calvin cycle. This reaction, however, is not the reverse of any reaction in glycolysis; the formation of fructose-1,6-bisphosphate that occurred there (p. 127) was at the expense of ATP and catalysed by a kinase, whereas the cleavage of fructose-1,6-bisphosphate is an hydrolysis and catalysed by a phosphatase.

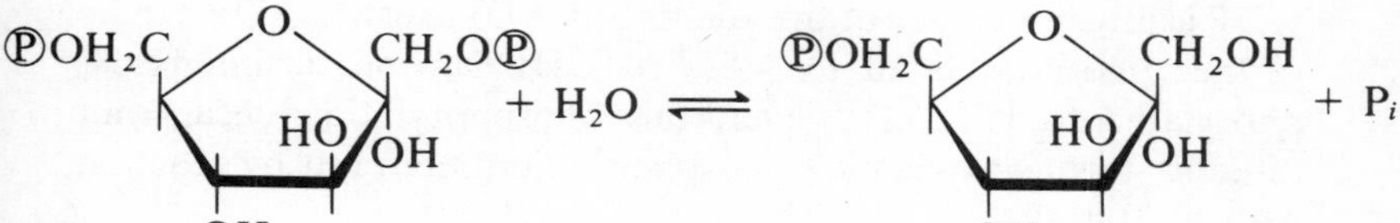

$$(22)$$

fructose-1,6-bisphosphate   fructose-6-phosphate

Owing to the presence of hexose phosphate isomerase, fructose-6-phosphate is converted to glucose-6-phosphate (see p. 126). Glucose-6-phosphate is in turn converted to glucose-1-phosphate by the action of a mutase. The first of these reactions is the reverse of reaction (2) of glycolysis (p. 127), and the second has been mentioned in the discussion of glycogen breakdown (reaction (15), p. 131). In the synthesis of glycogen from *glucose*, for example in the liver after absorption of glucose from the gut, glucose is first phosphorylated by the hexokinase reaction (p. 126). The resulting glucose-6-phosphate is then converted to glucose-1-phosphate by the mutase.

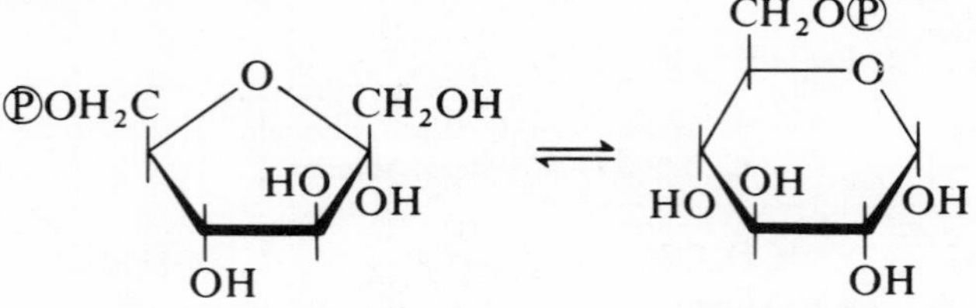

$$\text{fructose-6-phosphate} \rightleftharpoons \text{glucose-6-phosphate} \tag{23}$$

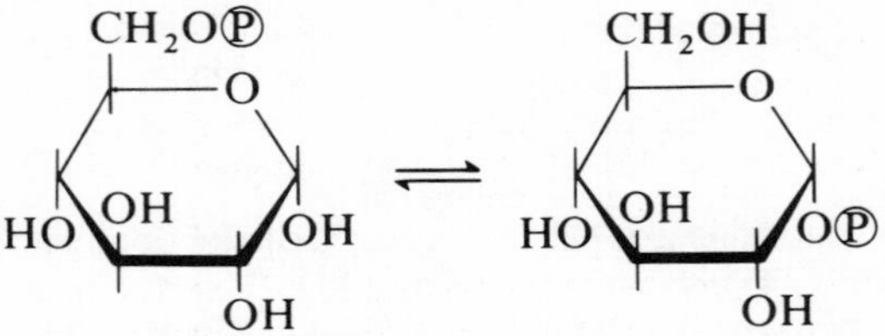

$$\text{glucose-6-phosphate} \rightleftharpoons \text{glucose-1-phosphate} \tag{24}$$

We have now arrived at the reactions that lead specifically to the synthesis of polysaccharides. In order to polymerize glucose, it is first necessary to convert the sugar into a form that has a high potential for transfer and can easily be donated to an acceptor molecule in the synthesis of the polymer.

Activation of the glucose molecule is brought about by attaching to it uridine diphosphate (UDP). UDP is analogous to ADP, in that it consists of a base, ribose and two phosphate groups; the base in UDP is uracil, in place of the adenine of ADP. Just as ADP can be further phosphorylated to ATP, so UDP can be further phosphorylated to UTP. However, this phosphorylation occurs not directly (through oxidative phosphorylation, etc.) but by reaction with ATP itself:

$$\text{ATP} + \text{UDP} \rightleftharpoons \text{ADP} + \text{UTP}. \tag{25}$$

UTP can now react with glucose-1-phosphate, the product of reaction (24), to form UDP-glucose.

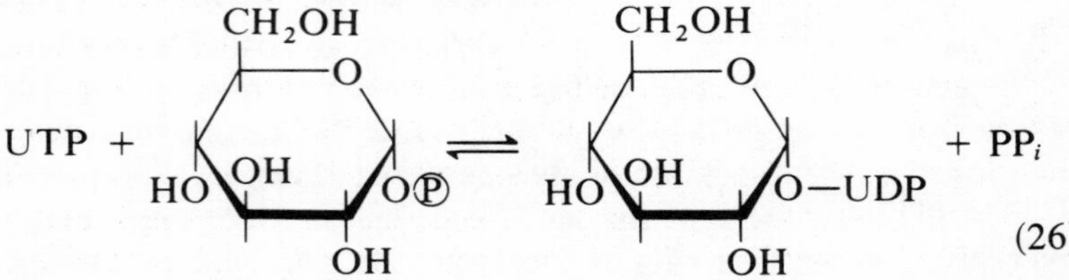

$$\text{UTP} + \text{glucose-1-phosphate} \rightleftharpoons \text{UDP-glucose} + \text{PP}_i \text{ (pyrophosphate)} \tag{26}$$

This reaction, catalysed by a pyrophosphorylase, has an equilibrium constant sufficiently near to one to allow net formation of product in either direction. However the rapid hydrolysis of pyrophosphate (p. 141) makes the reaction from right to left in practice impossible.

UDP-glucose contains a 'high-energy' bond between a phosphate residue and a glucose residue, and the glucose can therefore be readily donated to a suitable acceptor. In the synthesis of glucose polymers the acceptor is a *primer* consisting of a few glucose residues, and the glucose from UDP-glucose becomes transferred to this primer and attached by its C-1, thus lengthening the chain by one unit. The enzyme is a transferase, and UDP is liberated:

$$(glucose)_n + UDP\text{-glucose} \rightleftharpoons (glucose)_{n+1} + UDP. \qquad (27)$$

In this way polymers such as amylose (consisting of glucose residues joined by $1 \rightarrow 4\alpha$ linkages) (see p. 67) or cellulose (consisting of glucose residues joined by $1 \rightarrow 4\beta$ linkages) (see p. 63) can be built up. Glycogen, which has $1 \rightarrow 6\alpha$ linkages as well as $1 \rightarrow 4\alpha$ linkages, is made from amylose by a branching enzyme. This enzyme transfers small fragments from the end of the amylose chain to the 6 position of glucose residues.

We can now calculate the requirement for ATP in the synthesis of polysaccharide from lactic acid. Reactions (14), (15) and (18) use up one molecule of ATP each; so three equivalents of ATP are used for the synthesis of one molecule of triose phosphate from lactic acid, and therefore six molecules of ATP are used to form hexose monophosphate from two molecules of lactic acid. A further molecule of ATP is used up in reaction (25); thus in total seven ATP molecules are required for each glucose residue originating in lactic acid and polymerized into glycogen. If we start from oxaloacetic acid the requirement will be five ATP molecules since oxaloacetic acid can enter at reaction (15).

What we have said implies that any compound that can provide carbon atoms to the Krebs cycle can be used to synthesize polysaccharide. There is one important sense in which this implication is misleading. When the acetyl group of acetyl coenzyme A enters the Krebs cycle it reacts with oxaloacetic acid to yield citric acid (p. 135), but after one complete revolution of the cycle only oxaloacetic acid remains; the acetyl group has been completely oxidized. Thus acetyl coenzyme A does not add to the stock of intermediates in the Krebs cycle. There is no way in which the cycle can make use of the carbon atoms of the acetyl group for any reaction other than oxidation.

The chief source of acetyl coenzyme A is fatty acids, and it is for these reasons that the fatty acids that we discussed in Chapter 12 cannot be used to provide carbon to replenish the intermediates in the Krebs cycle. For exactly the same reasons fatty acids cannot be converted into carbohydrate. Although for other substances the Krebs cycle functions both as the terminal path of oxidation and as a crossroads of metabolic conversions, for fatty acids the cycle can act only as a crematorium.

16 Uses of ATP and NADPH₂ – the synthesis of fats

In Chapter 7 we explained that energy-yielding processes could give rise to ATP or to one of the reduced coenzymes $NADH_2$ or $NADPH_2$. During the breakdown of biological molecules to simpler compounds oxidation occurs and ATP is synthesized. During the synthesis of larger molecules from small ones reduction with $NADH_2$ or $NADPH_2$ occurs and ATP is used. In any given synthesis either the oxidation of $NADH_2$ to NAD (or $NADPH_2$ to NADP) or the breakdown of ATP to ADP may predominate. We saw in the last chapter that the synthesis of triose phosphate from carbon dioxide required both $NADPH_2$ and ATP, whereas the synthesis of polysaccharide from triose phosphate required a further supply of ATP but no further reduction. In this chapter we shall find that the synthesis of fatty acid from acetyl coenzyme A again requires both $NADPH_2$ and ATP.

We pointed out at the end of the last chapter that carbohydrate could not be synthesized from fatty acids. Fatty acids, on the other hand, can be synthesized from any compound that gives rise to acetyl coenzyme A. One of the chief sources of acetyl coenzyme A is, of course, carbohydrate. Thus fat can readily be synthesized from carbohydrate, which is every glutton's experience.

The key reaction in fatty-acid synthesis is a rather unexpected one, the carboxylation of acetyl coenzyme A to malonyl coenzyme A. This reaction requires ATP and is catalysed by the enzyme acetyl coenzyme A carboxylase, which (like many carboxylases) requires the co-factor biotin (p. 27).

$$CH_3CO—SCoA + CO_2 + ATP \rightleftharpoons$$
$$COOHCH_2CO—SCoA + ADP + P_i. \qquad (1)$$

acetyl coenzyme A   malonyl coenzyme A

Because ATP is used up, the equilibrium constant of this reaction overwhelmingly favours formation of malonyl coenzyme A. We saw in the last chapter that the biosynthetic sequences leading to triose phosphate and to polysaccharide contain steps that are effectively irreversible; here too the biosynthetic sequence leading to fatty acid contains an 'irreversible' reaction.

The malonyl group from malonyl coenzyme A is now bound to a small protein called the 'acyl carrier protein' (ACP). The linkage between the malonyl group and the ACP is a thiol ester, similar to the linkage between the malonyl group and coenzyme A.

$$COOHCH_2CO-SCoA + ACPSH \rightleftharpoons$$
$$COOHCH_2CO-SACP + CoASH. \quad (2)$$

In an exactly similar reaction, the acetyl group from another molecule of acetyl coenzyme A is also transferred to an acyl carrier protein.

$$CH_3CO-SCoA + ACPSH \rightleftharpoons CH_3CO-SACP + CoASH. \quad (3)$$

It is now possible for the two units to condense together. This reaction occurs by an attack of the carboxyl carbon of acetyl ACP (marked with an asterisk) on the methyl carbon of malonyl ACP (marked with a). Carbon dioxide is eliminated.

$$CH_3C^*O-SACP + COOHC^aH_2CO-SACP \rightleftharpoons$$
$$ACPSH + CH_3C^*OC^aH_2CO-SACP + CO_2. \quad (4)$$
acetoacetyl-ACP

This carbon dioxide is that that was used in reaction (1) to synthesize malonyl coenzyme A from acetyl coenzyme A. In a sense reaction (4) is really a condensation of two acetyl groups, and the carbon dioxide that made one of the two a malonyl group has served a catalytic role in the condensation.

The acetoacetyl group formed in reaction (4) must now be reduced. The first step in this process is the formation of a –CHOH group from the carbonyl group.

$$CH_3COCH_2CO-SACP + NADPH_2 \rightleftharpoons$$
$$CH_3CHOHCH_2CO-SACP + NADP. \quad (5)$$

Next, water is eliminated, and the double bond that results is again reduced (we mentioned this latter reaction on p. 104).

$$CH_3CHOHCH_2CO-SACP \rightleftharpoons$$
$$CH_3CH=CHCO-SACP + H_2O. \quad (6)$$

$$CH_3CH=CHCO-SACP + NADPH_2 \rightleftharpoons$$
$$CH_3CH_2CH_2CO-SACP + NADP. \quad (7)$$

In principle, these reactions are the reverse of the three reactions involved in the oxidation of fatty-acyl coenzyme A to β-ketoacyl coenzyme A (p. 141). There are, however, important differences. In the first place, each of the intermediates is tightly bound to its acyl carrier protein, and the intermediate–ACP complex is itself tightly bound to a complex containing the various enzyme activities necessary for these reactions. Secondly, whereas NAD is used as the acceptor of one pair of hydrogen atoms in the oxidation of fatty acids (the other pair being passed direct to a flavoprotein), the synthesis of fatty acids requires $NADPH_2$. (We recall that the principal source of $NADPH_2$ is the direct oxidation of glucose-6-phosphate (p. 145).) So while in principle it would be possible to oxidize fatty acids by a reversal of reactions (5), (6) and (7), in practice the degradative sequence and the synthetic sequence are quite distinct. Their separate identity is reinforced by the 'irreversible' steps in each sequence – the reactions catalysed by the thiokinase and β-keto thiolase in degradation, and the carboxylation reaction (1) in synthesis.

The reactions so far described have led to the production of butyryl-ACP, and thus the synthesis of a fatty-acyl group two carbon atoms longer than the original acetyl group has been achieved. It is now possible for another condensation to occur, similar to that in reaction (4). The butyryl group now takes the place of the acetyl group in reaction (4), and condenses with another malonyl group; carbon dioxide is again eliminated.

$$CH_3CH_2CH_2CO—SACP + COOHCH_2CO—SACP \rightleftharpoons$$

$$CH_3CH_2CH_2COCH_2CO—SACP + ACPSH + CO_2 \qquad (8)$$

The β-oxo-hexanoyl-ACP that results is now reduced by a repetition of reactions (5), (6) and (7).

$$CH_3CH_2CH_2COCH_2CO—SACP$$
$$\downarrow NADPH_2$$
$$CH_3CH_2CH_2CHOHCH_2CO—SACP$$
$$\downarrow$$
$$CH_3CH_2CH_2CH{=}CHCO—SACP$$
$$\downarrow NADPH_2$$
$$CH_3CH_2CH_2CH_2CH_2CO—SACP$$

hexanoyl-ACP

In this way long-chain fatty acids are built up. The essentials of each cycle of reactions are condensation of an acyl group with a malonyl group (each attached to its acyl carrier protein), followed by reduction with two molecules of $NADPH_2$. We can summarize the synthesis of stearoyl-ACP as follows.

$$CH_3CO—SACP + 8COOHCH_2CO—SACP + 16NADPH_2 \rightarrow$$

$$CH_3(CH_2)_{16}CO—SACP + 8CO_2 + 8ACPSH + 16NADP + 8H_2O.$$

This formulation emphasizes three facts. First the complex of intermediates, enzymes and acyl carrier proteins is a tight one (compare the complexes we mentioned on p. 39); and the cycle of condensation and reduction continues in this complex, without the liberation of intermediates, until the long-chain fatty-acyl group is completed. Secondly, all of the two-carbon units (except the first) that are condensed into fatty acids enter the sequence as malonyl groups. Thirdly, each malonyl group has to be synthesized in a reaction involving ATP and each has to be reduced by two molecules of $NADPH_2$. Another way of writing the overall reaction may make this last point even clearer.

$$9CH_3CO—SCoA + 8CO_2 + 8ATP + 16NADPH_2 \rightarrow$$

$$CH_3(CH_2)_{16}CO—SCoA + 8CO_2 + 8ADP + 8P_i + 16NADP$$
$$+ 8CoASH.$$

We must now consider how a triglyceride is synthesized. The starting compounds are fatty-acyl coenzyme A and glycerol-3-phosphate. Fatty-acyl coenzyme A is formed from fatty-acyl-ACP by a reversal of a reaction similar to (2) or (3).

$$CH_3(CH_2)_n CO—SACP + CoASH \rightleftharpoons$$

$$CH_3(CH_2)_n CO—SCoA + ACP.$$

Glycerol-3-phosphate can be formed by phosphorylation of glycerol (p. 131),

$$\begin{array}{c} CH_2OH \\ | \\ CHOH + ATP \rightleftharpoons \\ | \\ CH_2OH \\ \text{glycerol} \end{array} \quad \begin{array}{c} CH_2OH \\ | \\ CHOH + ADP \\ | \\ CH_2O\text{\textcircled{P}} \\ \text{glycerol-3-} \\ \text{phosphate} \end{array}$$

or by reduction of dihydroxyacetone phosphate.

$$\begin{array}{c} CH_2OH \\ | \\ CO \quad + NADH_2 \rightleftharpoons \\ | \\ CH_2O\text{\textcircled{P}} \\ \text{dihydroxyacetone} \\ \text{phosphate} \end{array} \quad \begin{array}{c} CH_2OH \\ | \\ CHOH + NAD \\ | \\ CH_2O\text{\textcircled{P}} \\ \text{glycerol-3-} \\ \text{phosphate} \end{array}$$

Two molecules of fatty-acyl coenzyme A can react with one molecule of glycerol-3-phosphate to give a compound known as phosphatidic acid (see p. 80).

$$\begin{array}{l} CH_3(CH_2)_nCO{-}SCoA \\ \quad + \\ CH_3(CH_2)_mCO{-}SCoA \\ \quad + \\ CH_2OH \\ | \\ CHOH \\ | \\ CH_2O\textcircled{P} \end{array} \rightleftharpoons \begin{array}{l} CH_2OCO(CH_2)_nCH_3 \\ | \\ CHOCO(CH_2)_mCH_3 \\ | \\ CH_2O\textcircled{P} \\ \text{phosphatidic acid} \end{array} + 2CoASH$$

Phosphatidic acid can now be hydrolysed to yield a diglyceride, and this will react with a third molecule of fatty acyl coenzyme A.

$$\begin{array}{l} CH_2OCO(CH_2)_nCH_3 \\ | \\ CHOCO(CH_2)_mCH_3 \\ | \\ CH_2O\textcircled{P} \end{array} + H_2O \rightleftharpoons \begin{array}{l} CH_2OCO(CH_2)_nCH_3 \\ | \\ CHOCO(CH_2)_mCH_3 \\ | \\ CH_2OH \\ \text{diglyceride} \end{array} + P_i$$

$$\begin{array}{l} CH_2OCO(CH_2)_nCH_3 \\ | \\ CHOCO(CH_2)_mCH_3 \\ | \\ CH_2OH \\ \quad + \\ CH_3(CH_2)_lCO{-}SCoA \end{array} \rightleftharpoons \begin{array}{l} CH_2OCO(CH_2)_nCH_3 \\ | \\ CHOCO(CH_2)_mCH_3 \\ | \\ CH_2OCO(CH_2)_lCH_3 \\ \text{triglyceride} \end{array} + CoASH$$

The final product of this sequence is thus triglyceride.

17 Amino-acid metabolism

In the chapters that described the main sources of ATP and reducing power we made no mention of the use of nitrogen-containing substances as substrates. By contrast in Chapters 3–6 we stressed that the nitrogen-containing macromolecules proteins and nucleic acids possess great specificity, and that proteins in particular play extremely diverse roles that depend on their specific structure. The chief biochemical importance of amino acids, then, is not so much that they can be broken down to provide ATP, but rather that they are the constituents of proteins; so that it is appropriate to concentrate in this chapter chiefly on the *synthesis* of amino acids. However, amino acids can, in fact, serve as minor sources of ATP, so we shall also touch briefly on this aspect of their metabolism.

The most important fact that we wish to stress about amino-acid metabolism is that it does not comprise a completely separate set of reactions unconnected with the sequences that we have described in previous chapters. The metabolism of amino acids is closely linked with that of carbohydrates and fats. In particular there is one reaction that is central to amino-acid metabolism, providing a link between amino acids and the Krebs cycle.

$$
\begin{array}{l}
\text{COOH} \\
|\\
\text{CO} \\
|\\
\text{CH}_2 \quad + \text{NH}_3 + \quad
\begin{array}{c}
\text{NADH}_2 \\
or \\
\text{NADPH}_2
\end{array}
\rightleftharpoons
\begin{array}{l}
\text{COOH} \\
|\\
\text{CHNH}_2 \\
|\\
\text{CH}_2 \quad + \text{H}_2\text{O} + \quad
\begin{array}{c}
\text{NAD} \\
or \\
\text{NADP}
\end{array}
\end{array} \\
|\\
\text{CH}_2 \\
|\\
\text{COOH} \\
\alpha\text{-oxoglutaric acid}
\end{array}
\qquad (1)
$$

α-oxoglutaric acid   glutamic acid

In this reaction α-oxoglutaric acid is reductively aminated to yield glutamic acid, the hydrogen atoms necessary for the reduction being provided by one of the reduced coenzymes. This reaction can be reversed, and the back reaction provides a means of oxidizing glutamic acid with the liberation of ammonia. The enzyme that catalyses the equilibrium is called glutamate dehydrogenase.

172

The great importance of this reaction is that in many organisms it is the sole means by which inorganic nitrogen can be fixed into a non-nitrogenous organic molecule to form an α-amino group. We shall see (p. 190) that there is a separate reaction for the fixation of ammonia to form carbamyl phosphate and thus pyrimidines; and also a reaction for the fixation of another molecule of ammonia into glutamic acid to form glutamine and thus purines. But the *de novo* synthesis of amino acids depends almost entirely on the synthesis of glutamic acid from α-oxoglutaric acid, since the amino group, once introduced into glutamic acid, can become transferred to yield other amino acids.

This type of transfer reaction is called a transamination, and it can be represented as follows.

$$
\begin{array}{llll}
\text{COOH} & & \text{COOH} & \\
| & & | & \\
\text{CHNH}_2 & \text{R} & \text{CO} & \text{R} \\
| & | & | & | \\
\text{CH}_2 & + \ \text{CO} & \rightleftharpoons \ \text{CH}_2 & + \ \text{CHNH}_2 \\
| & | & | & | \\
\text{CH}_2 & \text{COOH} & \text{CH}_2 & \text{COOH} \\
| & & | & \\
\text{COOH} & & \text{COOH} &
\end{array} \qquad (2)
$$

Its essential feature is an exchange of an α-amino group with an α-keto group, catalysed by an enzyme called a transaminase. Pyridoxal phosphate is required as coenzyme; this is formed from a vitamin (see p. 27). Much the commonest transaminations (in some organisms the *only* transaminations) are those in which one of the reactants is glutamic acid. The other reactant may be any one of a number of α-keto acids; the specificity of the enzyme determines which α-keto acid receives the amino group from glutamic acid.

One of the two products of the transamination is, of course, the amino acid corresponding to the keto acid that participated in the reaction. The other product is α-oxoglutaric acid; and this can now once more be converted to glutamic acid by reaction (1). Thus the couple glutamic acid–α-oxoglutaric acid can function catalytically in the synthesis of an amino acid, indirectly fixing ammonia into the α-keto acid,

$$
\begin{array}{c}
\qquad \text{glutamic acid} \qquad\qquad \text{RCOCOOH} \\
\\
\text{ammonia} \qquad \alpha\text{-oxoglutaric acid} \qquad \text{RCHNH}_2\text{COOH}
\end{array} \qquad (3)
$$

and provided that the Krebs cycle is functioning normally (in other words if there is adequate metabolism of carbohydrate, see p. 143) the supply of α-oxoglutaric acid will be sufficient for this purpose.

So far we have concentrated on the synthesis of glutamic acid by fixation of ammonia and the use of the amino group thus formed in the synthesis of other amino acids. But reaction (2), like reaction (1), is reversible; thus the couple glutamic acid–α-oxoglutaric acid can equally well function in the deamination of amino acids.

$$\text{ammonia} \overset{\displaystyle \text{glutamic acid}}{\underset{\displaystyle \alpha\text{-oxoglutaric acid}}{\rightleftharpoons}} \quad \overset{\displaystyle \text{RCOCOOH}}{\underset{\displaystyle \text{RCHNH}_2\text{COOH}}{}} \tag{4}$$

These reactions indirectly remove ammonia from an amino acid, leaving the corresponding α-keto acid which can be further metabolized (see below).

A further use of the glutamic acid–α-oxoglutaric acid couple is in the transfer of an amino group from one amino acid to another without the intervention of free ammonia. This reaction is useful in circumstances when one amino acid is relatively abundant and some others in short supply, and it requires the use of two transaminases.

$$\overset{\displaystyle \text{R'COCOOH}}{\underset{\displaystyle \text{R'CHNH}_2\text{COOH}}{}} \quad \overset{\displaystyle \text{glutamic acid}}{\underset{\displaystyle \alpha\text{-oxoglutaric acid}}{\rightleftharpoons}} \quad \overset{\displaystyle \text{RCOCOOH}}{\underset{\displaystyle \text{RCHNH}_2\text{COOH}}{}} \tag{5}$$

By the use of these various reactions, organisms can convert any of the α-keto acids formed in the common metabolic reaction sequences into the corresponding amino acids. Thus alanine can be produced from the pyruvic acid that is an intermediate in the Embden–Meyerhof pathway,

$$\overset{\displaystyle \text{CH}_3\text{COCOOH}}{\underset{\displaystyle \text{CH}_3\text{CHNH}_2\text{COOH}}{}} \quad \overset{\displaystyle \text{COOHCHNH}_2\text{CH}_2\text{CH}_2\text{COOH}}{\underset{\displaystyle \text{COOHCOCH}_2\text{CH}_2\text{COOH}}{}} \tag{6}$$

and aspartic acid can be produced from the oxaloacetic acid that is an intermediate in the Krebs cycle.

$$\overset{\displaystyle \text{COOHCOCH}_2\text{COOH}}{\underset{\displaystyle \text{COOHCHNH}_2\text{CH}_2\text{COOH}}{}} \quad \overset{\displaystyle \text{COOHCHNH}_2\text{CH}_2\text{CH}_2\text{COOH}}{\underset{\displaystyle \text{COOHCOCH}_2\text{CH}_2\text{COOH}}{}} \tag{7}$$

Similarly an amino acid can be formed from a keto acid which, though not itself a common metabolic intermediate, can be readily made from a starting material that is on a normal metabolic pathway. An example is the synthesis of serine. Glyceric acid-3-phosphate (p. 128) is oxidized, then transaminated, and finally dephosphorylated by the specific enzyme phosphoserine phosphatase.

$$
\begin{array}{c}
\text{NAD} \quad \text{NADH}_2 \\
\begin{array}{l}
\text{COOH} \\
| \\
\text{CHOH} \\
| \\
\text{CH}_2\text{O}\textcircled{P}
\end{array}
\xrightarrow{}
\begin{array}{l}
\text{COOH} \\
| \\
\text{CO} \\
| \\
\text{CH}_2\text{O}\textcircled{P}
\end{array}
\rightleftharpoons
\begin{array}{l}
\text{COOH} \\
| \\
\text{CHNH}_2 \\
| \\
\text{CH}_2\text{O}\textcircled{P}
\end{array}
\rightleftharpoons
\begin{array}{l}
\text{COOH} \\
| \\
\text{CHNH}_2 \\
| \\
\text{CH}_2\text{OH}
\end{array}
\end{array} \tag{8}
$$

Many of the reaction sequences that lead to the formation of amino acids are rather complicated; but the transformations involved are generally similar to those that we have already outlined. A few include special reactions – for example the synthesis of the rings of phenylalanine and tyrosine, and of histidine – but we shall not consider these here. The point that we wish to establish is that there are two ways in which the synthesis of amino acids is connected to the metabolic sequences that we have previously described. First there is the provision of the amino group by transamination involving the glutamic acid–α-oxoglutaric acid couple. Secondly, as we have seen in reactions (6), (7) and (8), there is the provision of the carbon atoms that form the backbone of the molecule.

Two of the most important starting compounds from which the backbones of the amino acids are formed are aspartic acid and glutamic acid. Aspartic acid can give rise to methionine, lysine and threonine; glutamic acid can give rise to proline and arginine. (Note that this use of glutamic acid for providing the carbon atoms of proline and arginine is separate from its function of donating the amino group in transaminations.) Now aspartic acid is produced by transamination of oxaloacetic acid, and glutamic acid (either directly or by transamination) from α-oxoglutaric acid. These reactions, then, represent major uses of carbon atoms from intermediates in the Krebs cycle (see Fig. 9.3), and are part of the reason why the carboxylation reaction for replenishing oxaloacetic acid (p. 139) is so essential.

Another compound formed from glutamic acid is glutamine, which is synthesized by fixing ammonia into the glutamic acid molecule.

$$
\begin{array}{l}
\text{COOH} \\
| \\
\text{CHNH}_2 \\
| \\
\text{CH}_2 \quad +\text{NH}_3+\text{ATP} \rightleftharpoons \\
| \\
\text{CH}_2 \\
| \\
\text{COOH}
\end{array}
\begin{array}{l}
\text{COOH} \\
| \\
\text{CHNH}_2 \\
| \\
\text{CH}_2 \quad +\text{ADP}+\text{P}_i \\
| \\
\text{CH}_2 \\
| \\
\text{CONH}_2
\end{array}
$$

glutamic acid glutamine

Glutamine is an amino acid that is a constituent of proteins (Table 3.1), and it also takes part in some reactions in intermediary metabolism as a donor of amino groups (see, for example, p. 191). In some organisms glutamine and α-oxoglutaric acid can react to yield two molecules of glutamic acid.

$$
\begin{array}{c}
\text{COOH} \\
| \\
\text{CHNH}_2 \\
| \\
\text{CH}_2 \\
| \\
\text{CH}_2 \\
| \\
\text{CONH}_2 \\
\text{glutamine}
\end{array}
+
\begin{array}{c}
\text{COOH} \\
| \\
\text{CO} \\
| \\
\text{CH}_2 \\
| \\
\text{CH}_2 \\
| \\
\text{COOH} \\
\alpha\text{-oxoglutaric} \\
\text{acid}
\end{array}
+ \text{NADPH}_2 \rightleftharpoons 2
\begin{array}{c}
\text{COOH} \\
| \\
\text{CHNH}_2 \\
| \\
\text{CH}_2 \\
| \\
\text{CH}_2 \\
| \\
\text{COOH} \\
2 \times \text{glutamic acid}
\end{array}
+ \text{NADP}
$$

What we have said in the last few pages may imply that all amino acids can be readily synthesized from starting materials that are present among the intermediates of carbohydrate and fat metabolism. This is true only of some organisms. Most animals, for example, are unable to synthesize all twenty of the amino acids; as a rough rule, most animals seem to be able to synthesize about half the total number. The remaining amino acids have to be provided in the diet (see p. 29), and these are known as essential amino acids (the precise list of essential amino acids differs slightly from one animal species to another). As these are needed in substrate quantities for the synthesis of protein they are readily distinguishable from the vitamins that are needed in catalytic quantities to act as co-factors. An amino acid is essential simply because the animal is unable to carry out one of the enzymic steps needed for its synthesis – generally one of the enzymic steps needed for the synthesis of the carbon-atom backbone. Thus, for example, although in most plants and many micro-organisms aspartic acid can (as we have just mentioned) be converted to methionine, lysine and threonine, in most higher animals all of these three amino acids are essential.

Reaction (5) above showed how one amino acid could donate its amino group, via glutamic acid, to form another amino acid. This conversion, however, is possible only if the appropriate keto acid (RCOCOOH in reaction (5)) is available. A required amino acid cannot be synthesized if the corresponding α-keto acid is not available, however much of the donor amino acid (R'CHNH$_2$COOH) is present. For this reason the nutritional value of a diet to an animal is not dependent merely on the quantity of protein it contains but also on the amino-acid composition of the protein. An animal can make use of proteins only to the extent that they satisfy its requirement of essential amino acids. If the proteins are lacking in an essential amino acid then the other amino acids present are of no use in satisfying nutritional requirements, and their nitrogen will be split off as ammonia (see reaction (4)). Since proteins in the diet never contain amino acids in precisely the proportions that the animal requires, there is always some wastage of amino acids and a corresponding release of their deamination products, α-keto acids.

This release of α-keto acids provides substrates that act (as we remarked at the beginning of this chapter) as minor sources of fuel. We do not intend to describe in detail the way in which the twenty amino acids can be broken down to provide ATP, but it is possible to outline certain general principles.

Since the deamination product of an amino acid is an α-keto acid, we can expect that this will be treated in the way that is usual for breakdown of α-keto acids. We have seen (p. 134) that the α-keto acid that is an important product of metabolism of carbohydrates, pyruvic acid, is broken down by oxidative decarboxylation and yields acetyl coenzyme A. In an analogous way the α-keto acid that is an intermediate in the Krebs cycle, α-oxoglutaric acid, is broken down by oxidative decarboxylation and yields succinyl coenzyme A. Although the α-keto acids derived from amino acids do not invariably undergo oxidative decarboxylation they do so quite generally. Again, as these two examples show, the product of oxidative decarboxylation of an α-keto acid is an acyl coenzyme A; and we may reasonably expect that this will be metabolized in the way that we described for acyl coenzyme A derivatives formed from fatty acids (Chapter 12).

The terminal product of this kind of metabolic process is normally either acetyl coenzyme A on the one hand, or a Krebs cycle intermediate or pyruvic acid on the other hand. (Pyruvic acid is grouped with the Krebs cycle intermediates because by carboxylation it can give rise to oxaloacetic acid (p. 139), whereas acetyl coenzyme A cannot be converted to a Krebs cycle intermediate.) The distinction between amino acids that are broken down to acetyl coenzyme A, and those that are broken down to a Krebs cycle intermediate or pyruvic acid, was the basis of an old classification of the amino acids as 'ketogenic' or 'glucogenic'. This classification relied on the fact that in a diabetic animal acetyl coenzyme A exacerbates ketosis by increasing the concentration of

acetoacetyl coenzyme A (see p. 143), whereas Krebs cycle intermediates can be converted to glucose (see pp. 161–4). The distinction is used less than it was, since we can now trace the entire
metabolic sequence of the degradation of amino acids rather than
being aware of no more than their end products.

Interlude. We have now covered in some detail most of the pathways
that we skimmed over in Chapter 9. (We have still to consider the
synthesis of the purine and pyrimidine nucleotides, which we shall
come to in Chapter 19.) This, then, might be a good moment for the
reader to refer again to the figures in Chapter 9 which summarize
the main pathways of intermediary metabolism, and to see how the
principles that we outlined in that chapter have been worked out in
the reaction sequences discussed in Chapters 10–17.

Activation of amino acids

At the beginning of this chapter we remarked that the chief biochemical importance of amino acids is the fact that they are constituents of proteins. We shall now start to see how proteins are in fact
synthesized from amino acids. In Chapter 15 we showed how, in the
synthesis of polysaccharide, glucose-1-phosphate has first to be
converted to uridine diphosphate glucose before polymerization.
We find that an analogous activation is necessary before amino
acids can form peptide bonds; the reaction for an amino acid is the
formation of the high-energy compound aminoacyl-AMP.

$$RCHNH_2COOH + ATP \rightleftharpoons RCHNH_2CO{-}AMP + PP_i.$$

The equilibrium constant of this reaction is near to one; but as we
have seen (p. 141) the fact that pyrophosphate is produced ensures
that in practice the reverse reaction to synthesize ATP does not
occur.

Aminoacyl-AMP contains a 'high-energy' bond and is analogous
to uridine diphosphate glucose (p. 164). One might therefore expect
that aminoacyl-AMP would be capable of condensing to form a
polyamino acid, splitting out AMP, in the same way as UDP-
glucose condenses to form a polyglucose, splitting out UDP. In fact,
such a reaction does not occur; and we must now consider why this
is.

The reason is exactly that which we gave in Chapter 2. Glucose
polymers are used for the storage of fuel in the form of carbohydrate, and they contain only one kind of monomeric residue. No
question of arrangement of residues arises; even the number of

residues in a polymer is not precisely determined. There is little specificity of structure and no need to ensure that one copy of a polymer molecule exactly resembles any other. Proteins, by contrast, are possessed of great specificity which allows them to perform all the functions that we outlined in Chapters 4 and 5. This specificity depends on an extremely precise ordering of the residues, and on an extreme reproducibility of the structure that ensures that every copy of the molecule is identical to every other.

Thus in the polymerization both of glucose and of amino acids, the residues must first be activated in the way that we have described. But in the synthesis of a glucose polymer no ordering of residues is required; so it is possible simply with the specificity of a single enzyme to ensure the formation of the correct linkage in the polymer (see p. 63). If proteins were random polymers of amino acids a similar condensation reaction would suffice; but since they are not, it is essential to find other means of ensuring that the precise ordering of amino-acid residues is maintained in protein synthesis.

How is this precision of ordering brought about? We know that the exact structure of many proteins is characteristic not just of a single individual of a species, but of the species as a whole, and that it is maintained through the generations (for example, it can be deduced that pig insulin has precisely the same structure now as it did many generations ago). So it seems that the ordering of amino-acid residues in a protein is dependent on the genetic material of the organism, and, indeed, that it is one of the functions of the genetic material to determine the ordering. Therefore before we can discuss how the precision of structure of proteins is achieved, we must first describe the functions and the mode of synthesis of the genetic material.

Section III
Molecular genetics and protein synthesis

18 The molecular basis of genetics

In Chapters 3–6 we outlined the ways in which structure and function are related in macromolecules. We did not describe in any detail one of the essential functions that macromolecules perform: the handling of *information*.

If one were entirely ignorant of the constitution of living matter at the molecular level one would hardly have guessed that the linear sequence of residues in a macromolecule could be used for the storage and transfer of information. The elaboration of molecules able to fulfil the role of information storage, and the development of mechanisms that bring about the transfer of information from one molecule to another, must rank as one of the great distinguishing features of living matter. It is of comparable importance with the development of macromolecules to overcome the limitations of the chemistry of free solution, with the extensive coupling between endergonic and exergonic reactions, and with the organization of metabolism into a large number of interlocking pathways. For this reason a whole section of this book has been devoted to informational macromolecules.

It is well known that DNA is chief among these macromolecules. In the rest of this chapter we will examine its role a little further and discuss its relation to genetics.

In many viruses, in all other micro-organisms and in all plants and animals, the genetic material is DNA. There is a wealth of experimental evidence that supports this statement, but we shall here mention only one kind of experiment. It is possible to extract and purify the DNA from a culture (the 'donor') of one particular species of bacteria and add it to another culture (the 'recipient'). Under certain conditions the recipient bacteria will take up the purified DNA and incorporate a piece of it in place of a piece of their own DNA. When this happens, the recipient bacteria may be found to have permanently acquired a characteristic of the donors – in other words, their genetic complement has been to that extent altered.

We remarked at the end of the last chapter that the genetic material appears to determine the structure of proteins, and the example that we gave was the control of the structure of insulin in the pig. If we wish to discover how the genetic material carries out this function, one obvious method is to see how *changes* in DNA are reflected in changes in the structure of proteins; but before we could find an altered insulin we might have to isolate the protein from thousands of separate pigs and in each case examine the structure of the molecule.

Another possible approach is to rely on the fact that enzymes are proteins of highly specific structure (see Chapter 5) and therefore that changes in their structure very frequently lead to loss of enzymic activity. So if we could find an organism that had lost a particular enzyme, we might be able to relate the loss to some change that had occurred in the DNA. However, with most organisms it is by no means easy to find individuals lacking particular enzymes; for if the enzyme is essential to metabolic activity we shall never find an individual without it, and conversely if the enzyme is inessential we shall not easily be able to recognize its loss.

We can overcome these difficulties by working with micro-organisms, because with micro-organisms we can alter conditions in such a way as to compensate for the loss of an enzyme. Suppose, for example, that a bacterium that is normally capable of producing serine by the pathway that we have described on p. 000 suffers a change in its DNA that results in a loss of the enzyme phosphoserine phosphatase. Serine is a constituent of all the proteins that the organism makes, and consequently the bacterium will be unable to grow without it. If, however, we include serine in the cultivation medium the bacterium will grow quite normally and we shall be able to study it. (We may say in this case that serine has become an essential amino acid for the bacterium. This example suggests that the reason why some amino acids are essential (p. 29) for certain species of animals is that the DNA of these animals has, in the course of evolution, become changed in such a way that the enzymes necessary for their synthesis have been lost.)

A change in the structure of DNA is called a mutation, and an organism that carries a mutation is called a mutant; by contrast the 'normal' organism, in which the mutation occurred, is called the wild type. (This definition is not a rigorous one, since organisms are described as wild type largely by convention, but in practice difficulty seldom arises.) Genetics has for many years made use of mutations without being able to define the precise change that occurred when the DNA underwent mutation, but recently it has become possible to correlate some mutations in DNA directly with changes in the structure of proteins.

The basic technique used in genetic studies is to make *crosses* between genetically different strains of a species – either between one mutant strain and another or between wild-type and mutant strains. (There are various means of making crosses and examining the progeny that result, but we shall not go into them here.) The value of genetic experiments is that their results allow one to make *genetic maps*. Just as the map of a railway line reveals the order of the stations and indicates the distances between them, so a genetic map reveals the order of the mutations along the DNA molecule and gives some idea of the distances between them. For a number of reasons that need not concern us, it is particularly easy to make very detailed maps of the DNA in bacteria as compared with higher organisms. We shall now mention some of the results that have emerged from bacterial genetics; almost all of them are derived from studies of the two closely related species *Escherichia coli* and *Salmonella typhimurium*.

The first point is that if several, independently isolated strains are found that are all deficient in a single enzyme, the mutations almost always map close together. To continue with the example that we gave previously, it may be possible to isolate several dozen separate strains of *E. coli* that are deficient in phosphoserine phosphatase; and if these mutations are mapped it is extremely likely that they will all be found to lie very close to one another in a small segment of DNA. This fact suggests that that segment of DNA is responsible for determining the structure of phosphoserine phosphatase, and that mutations in any of a number of places in that segment lead to loss of the phosphatase. We can speak of such a segment of DNA as a *gene* – in this particular example the gene for phosphoserine phosphatase.

Mutations can be of several kinds. The simplest kind of mutation is one in which a base in the polydeoxyribonucleotide chain (see p. 53) is replaced by another base – for example adenine may be replaced by guanine (since the bases in DNA are complementary, it would be more accurate to say that the base pair adenine–thymine is replaced by the base pair guanine–cytosine). This kind of mutation is called base substitution. Another kind of mutation involves the loss of a whole stretch of nucleotides (perhaps even many hundred nucleotides) from the two strands of DNA; this is called deletion. Yet another kind is the insertion of a stretch of nucleotides that really belong elsewhere into the middle of a gene; this is called insertion.

We do not know in detail how these changes in the structure of DNA occur. They happen at a low rate 'spontaneously', i.e. without experimental intervention, but the rate of their occurrence can be greatly increased by treating bacteria either with certain chemicals or with some kinds of radiation. The mode of action of some of these experimental treatments is known, and we can imagine that the spontaneous mutations occur in analogous ways – e.g. because all living organisms are subject to some background radiation.

In mutants deficient in a particular enzyme that have arisen through base substitution it is often possible to find in the cell a protein that is extremely similar to the lost enzyme. Sometimes it has been possible to prove that this new protein is in fact identical to the lost enzyme except for a change in a single amino acid residue. Moreover, in some cases in which a whole set of mutants has been discovered, all altered in a single gene, each mutant has been found to contain a distinct protein that is related, by a change in a single amino-acid residue, to the original enzyme. If now these mutants are used in genetic crosses it is possible to map the position of their mutations within the single gene; and from studies of the mutant proteins it is possible to construct a map of amino-acid changes within the single protein. These two maps correspond extremely well. The result shows that the *linear sequence* of the DNA is precisely related to the *linear sequence* of the protein, in other words that the ordering of bases in the DNA precisely corresponds to (and therefore must determine) the ordering of amino acids in the protein. This determining of the order of amino acids is often called 'coding'; we say that a sequence of bases in the DNA 'codes for' a sequence of amino acids in the protein. In Chapter 20 we shall see how this coding actually works.

Meanwhile we shall mention one further result of genetic studies with bacteria. We have already seen that mutations that lead to the loss of a single enzyme generally lie very close together. In bacteria, genes that determine the enzymes of one *pathway* often lie close together too. For example, there are three enzymes that are specifically needed for the fermentation of galactose and are not needed for other fermentations. The structures of these three enzymes are determined by three genes, and these are contiguous on the DNA of *E. coli*. A similar clustering of genes is often found for biosynthetic sequences. A striking example is the pathway of synthesis of histidine, which involves ten specific enzymes. The genes that determine the structure of these form an uninterrupted stretch of bacterial DNA. Clustering of this sort is not found universally even in bacteria (for instance the genes that determine the structures of the enzymes that synthesize arginine are scattered around the DNA of *E. coli*), but where it does occur it seems to be important in the control of enzyme synthesis (see Chapter 22).

So far in this chapter, we have discussed the results of genetic experiments largely in terms of bacteria, since much of the classical work that established the principles of molecular genetics was done with them. More recently techniques have been developed for studying the detailed genetic structure of higher organisms. Experiments with higher organisms have revealed some interesting phenomena that do not seem to occur in bacteria. For example, it has been discovered that a good deal of the DNA in higher organisms does not code for any protein. Still more remarkable, non-coding DNA sometimes appears in the middle of a gene, where it interrupts the sequence of bases that code for the amino acids of a single protein. (It follows that during the processes that lead to the synthesis of the protein there must be a mechanism for ensuring that, even though the chain of amino acids linked by peptide bonds is continuous, the non-coding DNA is not used to direct the insertion of amino acids.) We do not understand the significance of these findings, which have radically altered our views of the genetic organization of higher organisms.

To a large extent, these discoveries have resulted from the development of techniques for *genetic manipulation*. We referred earlier in this chapter to the classical technique of making crosses between genetically different strains of a species, and mentioned that its value was that it allowed one to make genetic maps. This has for many years been the basic technique of genetics: its particular feature is that it uses *biological* experiments that involve the whole of an organism's genetic material, so that inferences can be drawn about the position of mutations by examination of the progeny resulting from the cross. By contrast, the new techniques of genetic manipulation involve *chemical* or *enzymic* treatment of selected parts of an organism's genetic material. It is now possible to isolate a gene, or a cluster of genes, that is of particular interest; it is possible to modify the DNA that represents this genetic material by introducing a known mutation at a previously chosen place; it is possible to take genetic material from one organism and link it covalently to genetic material from another.

Perhaps the most striking genetic manipulation has been a development of this last technique: one can choose a particular gene from a higher organism, link it covalently to DNA that normally resides in a micro-organism, and put the resulting 'mixed' molecule back into the micro-organism. In some circumstances the cell will now replicate this molecule (see next chapter), with the result that all its descendants will contain copies of the higher organism's gene. The importance of this technique is that, since micro-organisms can grow and divide extremely rapidly, one will obtain an enormous amplification in quantity of the chosen gene, which in turn might be used to synthesize large quantities of the product of that gene. If you now imagine that the gene in question is that for a human hormone of medical value that has hitherto been available only in minute amounts at great expense, you will gain some idea of why genetic manipulation is of such practical importance, quite aside from the outstanding contribution it has made to our knowledge of the genetic structure of higher organisms.

19 Synthesis of DNA and RNA

The genetic material of a cell has two essential functions. One is to control the activities of the cell by specifying the structure of its components, especially proteins; we have seen in cursory outline in the last chapter, and shall examine in greater detail in the next chapter, how the linear sequence of nucleotides in DNA is related to the linear sequence of amino acids in proteins. The second function of the genetic material is to provide for its own exact replication, so that (barring the accident of mutation) each of the two daughter cells formed on division of the parental cell contains a precise copy of the genetic material of its parent. In this chapter we shall describe how the DNA replicates, in other words how DNA is synthesized in the cell and how the synthesis is controlled in such a way that the structure of the new genetic material is exactly similar to that of the old. We shall also describe how RNA is synthesized, because that process has much in common with the synthesis of DNA, and because a description of RNA synthesis is necessary for a consideration (in the next chapter) of protein synthesis.

Both DNA and RNA are synthesized from nucleoside triphosphates – deoxyadenosine triphosphate, deoxyguanosine triphosphate, deoxycytidine triphosphate and deoxythymidine triphosphate are the precursors of DNA, and adenosine triphosphate, guanosine triphosphate, cytidine triphosphate and uridine triphosphate are the precursors of RNA. These nucleoside triphosphates are simply the nucleoside monophosphates that we have already mentioned when describing DNA and RNA (p. 53), but with each nucleotide carrying two extra phosphate groups. ATP is, of course, a familiar example of a nucleoside triphosphate, and the others are analogous in structure. As we have seen previously (p. 52), ATP and deoxyATP contain adenine, and GTP and deoxyGTP contain guanine; both adenine and guanine are purines. CTP and deoxyCTP contain cytosine, UTP contains uracil and deoxyTTP contains thymine; cytosine, uracil and thymine are all pyrimidines. Most organisms are able to synthesize both purines and pyrimidines, and we shall outline briefly their pathways of synthesis, and the formation of the nucleoside triphosphates, before showing how polymerization to form DNA and RNA occurs.

The synthesis of nucleotides

The pathways of synthesis of purines and pyrimidines are rather dissimilar, but one thing that they have in common is that each requires a separate reaction for the fixation of ammonia. We have already seen (p. 175) that the fixation of ammonia into glutamic acid yields glutamine, and glutamine, as we shall see below, is used in the synthesis of purines. The reaction for the fixation of ammonia that is involved in the synthesis of pyrimidines is the formation of carbamyl phosphate.

$$CO_2 + NH_3 + 2ATP \rightleftharpoons NH_2COO\textcircled{P} + 2ADP + P_i$$

(Carbamyl phosphate, in addition to its use in the synthesis of pyrimidines, is essential for the synthesis of urea, which is the compound that many animals (including ourselves) form in order to excrete surplus nitrogen.)

The pathway of purine synthesis is complicated and we shall not consider it in any detail. Perhaps its most interesting feature is that the purine rings are built up, piece by piece, on a ribose-5-phosphate foundation. The effect of this is that, when the purine rings are completed, the molecule is already in the form of a nucleotide – actually the molecule inosinic acid (IMP), from which AMP and GMP can be easily formed. We recall that ribose-5-phosphate is formed by the oxidation of glucose-6-phosphate followed by isomerization (pp. 145–6), and in discussing the importance of this reaction we pointed out that ribose is an essential constituent of nucleotides. In the first reaction of purine biosynthesis, ribose-5-phosphate receives a diphosphate group from ATP to yield 5-phosphoribosyl diphosphate.

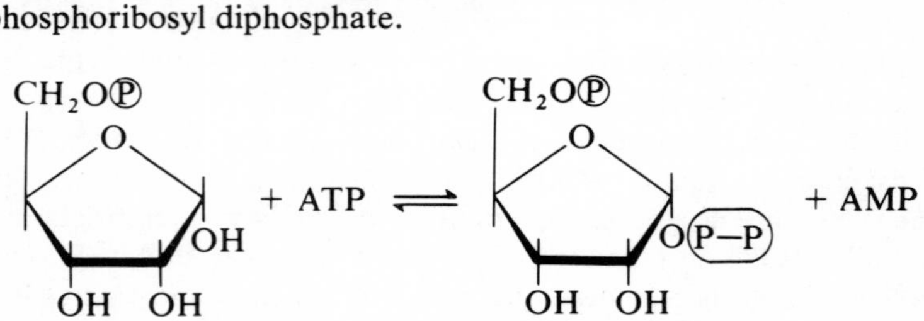

ribose-5-phosphate   5-phosphoribosyl diphosphate

This molecule then reacts with glutamine to give 5-phosphoribosylamine.

5-phosphoribosyl diphosphate — glutamine — 5-phosphoribosylamine — glutamic acid

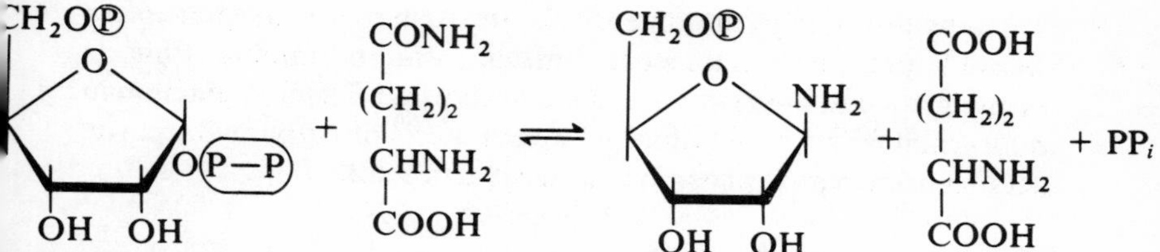

The nitrogen atom that is now attached to the 1-position of the ribose-5-phosphate remains in this position throughout the synthesis of the purine rings. During the synthesis several molecules react successively with the 5-phosphoribosylamine – including another molecule of glutamine, which provides one of the other nitrogen atoms of the purine. The final products are AMP and GMP, whose structures are now given again.

AMP — GMP

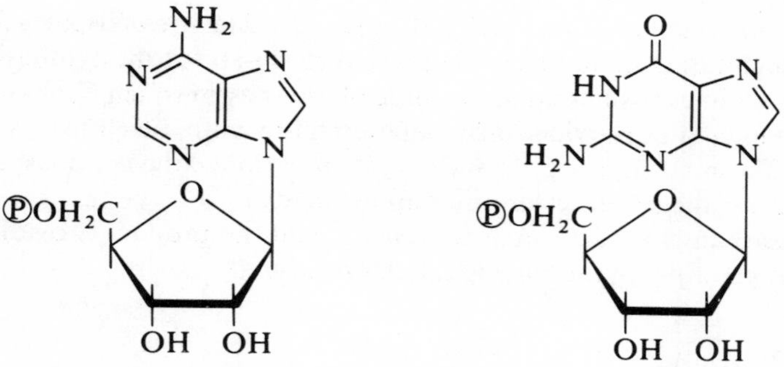

GMP can receive another phosphate group from ATP in a kinase-catalysed reaction:

$$\text{GMP} + \text{ATP} \rightleftharpoons \text{GDP} + \text{ADP}.$$

The nucleoside *di*phosphates can be converted to deoxynucleoside diphosphates in a reaction that involves reduction with $NADPH_2$. GDP, deoxyADP and deoxyGDP can all receive a third phosphate group from ATP in another kinase-catalysed reaction. This completes the synthesis of the four purine nucleoside triphosphates which, as we shall see below, are the substrates for the polymerase enzymes that catalyse the synthesis of DNA and RNA.

The synthesis of pyrimidines, by contrast, does not take place on a ribose-5-phosphate molecule. Instead, the pyrimidine ring is formed first and then converted to a nucleotide. The first reaction in the formation of pyrimidines is the synthesis of carbamyl aspartic acid from carbamyl phosphate and aspartic acid.

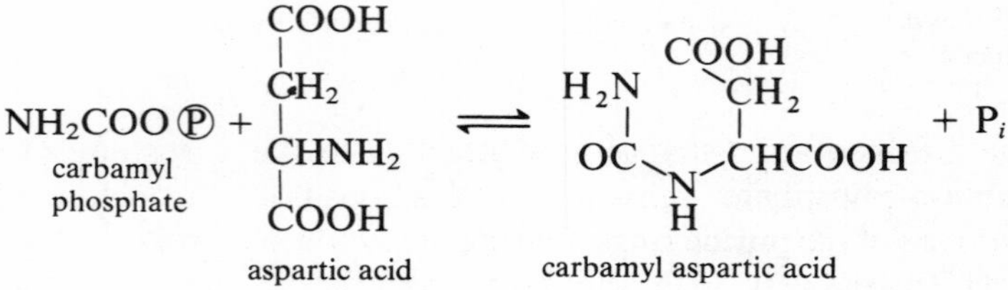

$$
\begin{array}{c}
\text{COOH} \\
| \\
\text{CH}_2 \\
\text{NH}_2\text{COO} \textcircled{P} + \quad | \\
\text{CHNH}_2 \\
| \\
\text{COOH}
\end{array}
\quad \rightleftharpoons \quad
\begin{array}{c}
\text{COOH} \\
\text{H}_2\text{N} \quad \text{CH}_2 \\
| \quad | \\
\text{OC}\diagdown_{\substack{\text{N} \\ \text{H}}}\diagup\text{CHCOOH}
\end{array}
\quad + \text{P}_i
$$

carbamyl phosphate　　　aspartic acid　　　　carbamyl aspartic acid

This reaction is catalysed by the enzyme aspartate transcarbamylase, which has been extensively studied as an example of an enzyme that is subject to 'end-product inhibition'. We shall discuss this phenomenon in some detail in Chapter 23. The use of aspartic acid (which derives from oxaloacetic acid (p. 174)) for the synthesis of pyrimidines is yet another example of the way in which Krebs cycle intermediates provide starting material for synthetic pathways (see p. 123 and compare p. 175, where we mentioned the use of aspartic acid for the synthesis of other amino acids).

Carbamyl aspartic acid is cyclized, and the product is oxidized. The resulting pyrimidine is called orotic acid.

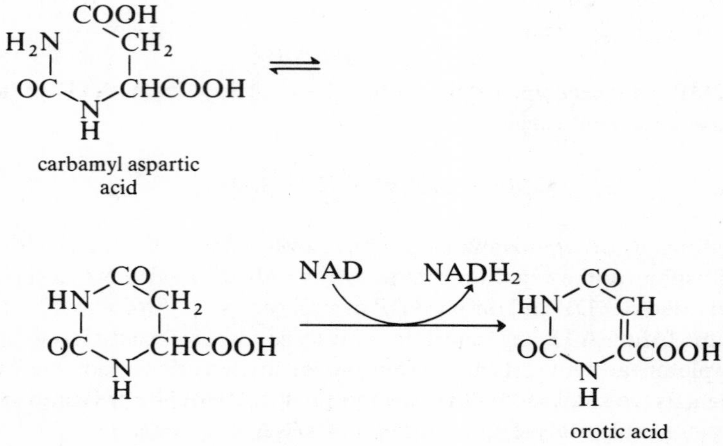

$$
\begin{array}{c}
\text{COOH} \\
\text{H}_2\text{N} \quad \text{CH}_2 \\
| \quad | \\
\text{OC}\diagdown_{\substack{\text{N} \\ \text{H}}}\diagup\text{CHCOOH}
\end{array}
\quad \rightleftharpoons
$$

carbamyl aspartic acid

$$
\begin{array}{c}
\text{CO} \\
\text{HN} \quad \text{CH}_2 \\
| \quad | \\
\text{OC}\diagdown_{\substack{\text{N} \\ \text{H}}}\diagup\text{CHCOOH}
\end{array}
\xrightarrow{\quad \text{NAD} \quad \text{NADH}_2 \quad}
\begin{array}{c}
\text{CO} \\
\text{HN} \quad \text{CH} \\
| \quad \| \\
\text{OC}\diagdown_{\substack{\text{N} \\ \text{H}}}\diagup\text{CCOOH}
\end{array}
$$

orotic acid

It is at this stage that ribose-5-phosphate is attached. The reactant, once again, is 5-phosphoribosyl diphosphate.

5-phosphoribosyl
diphosphate

orotic acid

$+ PP_i$

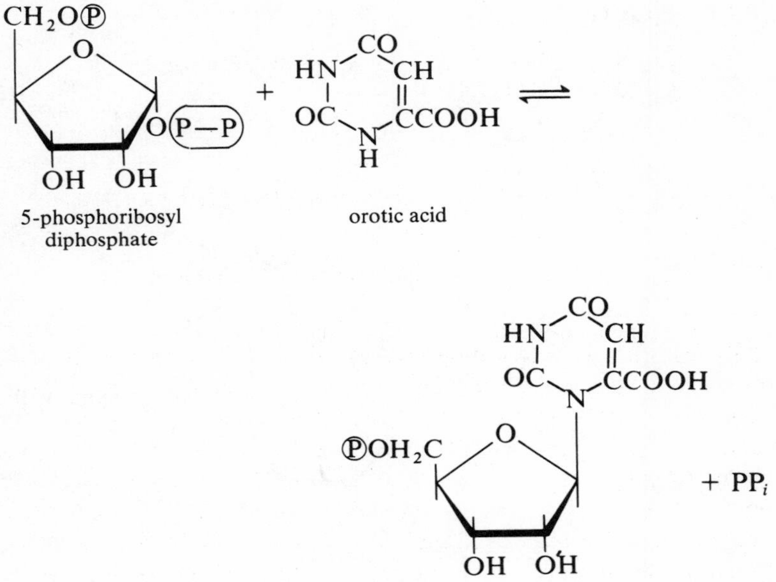

The resulting orotidine-5′-phosphate is decarboxylated to yield uridine-5′-phosphate (uridylic acid or UMP).

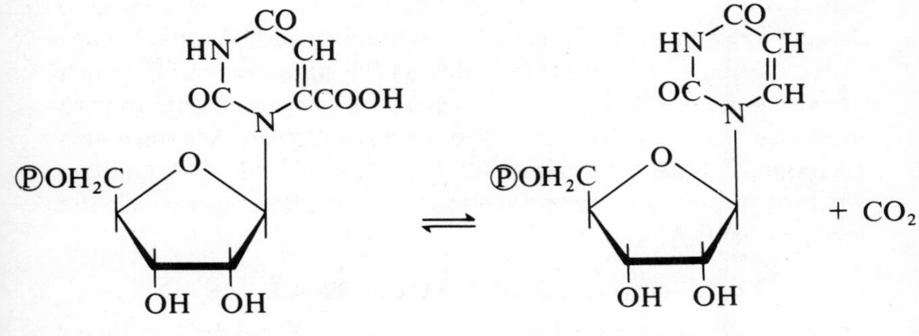

$+ CO_2$

UMP

By two successive reactions with ATP, UMP can be converted to UTP; and this, by reaction with ammonia, can give rise to CTP.

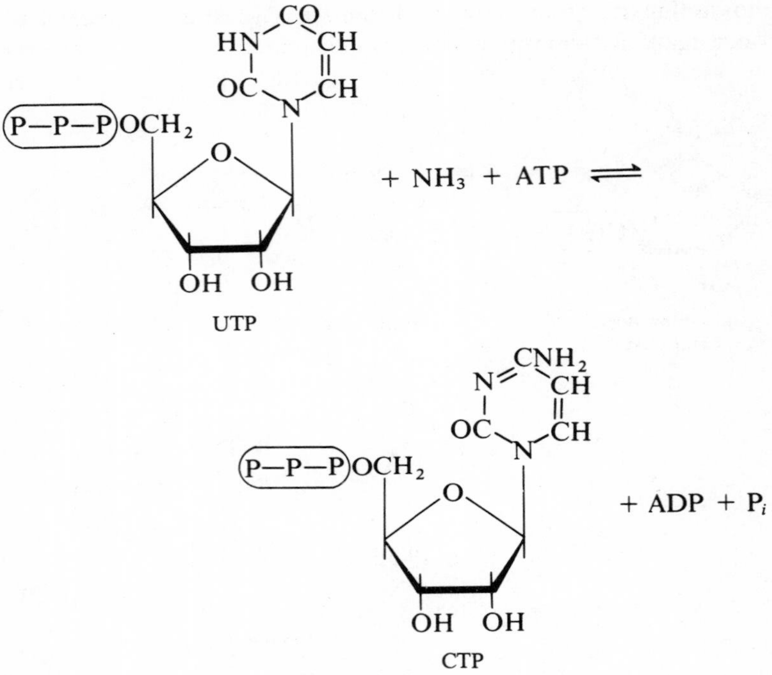

UTP

+ NH$_3$ + ATP $\rightleftharpoons$

CTP

+ ADP + P$_i$

CDP, which derives from CTP, can be reduced to deoxyCDP, in a reaction with NADPH$_2$ similar to that involved in the formation of deoxyADP and deoxyGDP (p. 191). DeoxyCDP leads to deoxyCTP by phosphorylation with ATP; alternatively, by a more complicated series of reactions, deoxyCDP can give rise to deoxyTTP. This rather sketchy description accounts for the formation of deoxyATP, deoxyGTP, deoxyCTP and deoxyTTP, which are the building blocks from which DNA is synthesized, and we must now consider the mechanism of this synthesis. We have also accounted for the synthesis of ATP, GTP, CTP and UTP; later we shall see (p. 200) how these nucleoside triphosphates are used in the synthesis of RNA.

Synthesis of DNA and RNA

We have previously referred in some detail (p. 53) to the hydrogen-bonding properties of the purine and pyrimidine bases. Let us now suppose that we wish to replicate a molecule of DNA, which we will call the parental molecule, and that we have a supply of all the deoxynucleotides. If we imagine the two strands of the parental molecule separating, we can see that each of the strands could act as a template which, owing to the specificity of hydrogen bonding between the bases, would ensure that a *complementary* sequence of nucleotides would become aligned opposite to itself. If these nucleotides now polymerized to form new strands of poly-nucleotides (daughter strands), we would have two double-stranded molecules of DNA. In each molecule the daughter strand would have the same sequence as the *opposite* parental strand (i.e. the parental strand in the other molecule). Consequently each molecule would consist of one parental strand and one daughter strand (that is, it would be half-old and half-new), and also each molecule would be identical in sequence to the original parental molecule.

An example will help to make this clear. If our parental double-stranded molecule has the sequence:

.... A-A-C-T-G-G-G-G-T-T-C-C-A-T-G
.... T-T-G-A-C-C-C-C-A-A-G-G-T-A-C

then the strands would separate to give:

.... A-A-C-T-G-G-G-G-T-T-C-C-A-T-G
and .... T-T-G-A-C-C-C-C-A-A-G-G-T-A-C.

Nucleotides would become aligned opposite these single strands in the following way (we use bold type to indicate the new nucleo-tides):

.... A-A-C-T-G-G-G-G-T-T-C-C-A-T-G
....**T T G A C C C C A A G G T A C**
and ....**A A C T G G G G T T C C A T G**
.... T-T-G-A-C-C-C-C-A-A-G-G-T-A-C

and these nucleotides would polymerize:

.... A-A-C-T-G-G-G-G-T-T-C-C-A-T-G
....**T-T-G-A-C-C-C-C-A-A-G-G-T-A-C**
and ....**A-A-C-T-G-G-G-G-T-T-C-C-A-T-G**
.... T-T-G-A-C-C-C-C-A-A-G-G-T-A-C.

These two DNA molecules are identical to one another, and they are also identical to the parental molecule. When the cell divides each daughter cell will receive one of the two molecules and will thus have exactly the same DNA as the parental cell.

This, then, in outline is how DNA is replicated. But the descrip-tion so far has referred only to the way in which specific base pairing is used to make two exact copies of the original DNA molecule. We have now to see how the replication occurs in terms of nucleotides rather than bases, and in particular to show how the polymerization reaction involves the deoxyribonucleoside triphosphates whose synthesis we have already described.

We have previously pointed out (p. 6) that if we consider carefully a single strand of polydeoxyribonucleotide we see that the two ends of the molecule are not identical. This difference between the ends is not a question of which bases are present, but a property of the sugar-phosphate backbone.

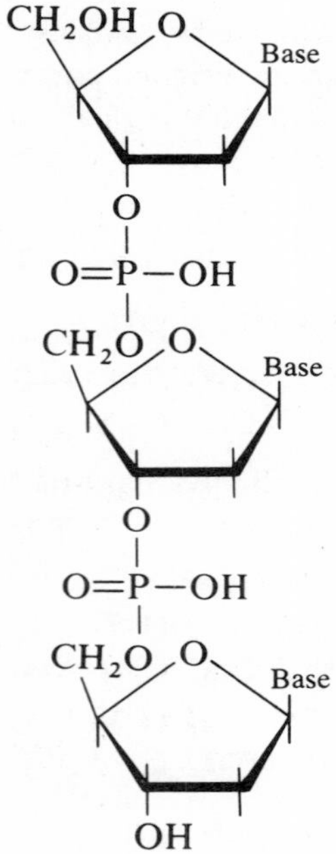

In this figure, we have drawn a tri-deoxyribonucleotide for the sake of simplicity, whereas in fact the strands of deoxyribonucleotides in DNA are extremely long. However, this miniature strand of DNA is sufficient to make clear the difference between the ends of the molecule. At the top we have a deoxyribose that has a free hydroxyl group on the 5'-carbon atom. At the bottom we have a deoxyribose that has a free hydroxyl group on the 3' end of the molecule. (In the middle we have a deoxyribose that has neither a free 5'-hydroxyl nor a free 3'-hydroxyl group; and in a real strand of DNA there would, of course, be thousands or even millions of residues of this form).

The route of synthesis of deoxyribonucleoside triphosphates that we described earlier in this chapter leads to deoxynucleosides with three phosphate groups attached to the 5′-carbon atom of deoxyribose (see pp. 191 and 194). In the synthesis of DNA a molecule of this kind reacts with the free 3′-hydroxyl end of a strand of polydeoxyribonucleotide.

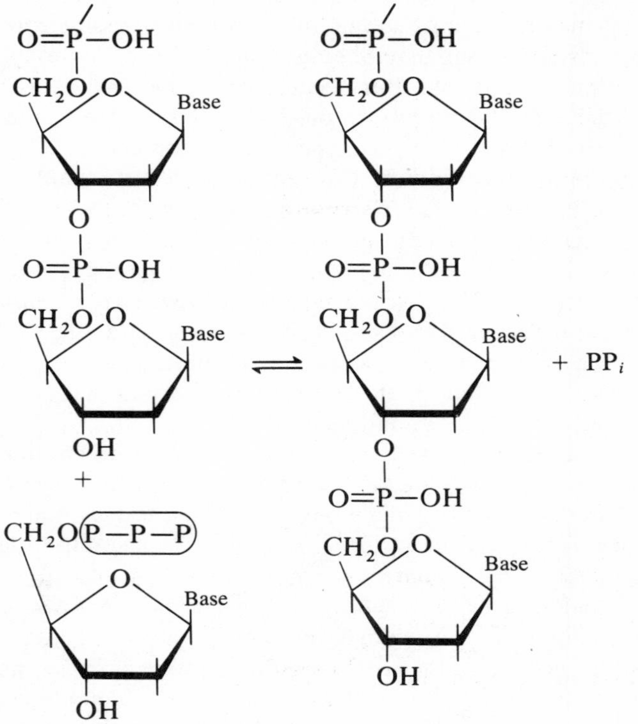

This reaction, catalysed by a DNA polymerase, produces inorganic pyrophosphate and therefore, for the reason that we discussed on p. 141, cannot be used in reverse for the degradation of DNA. Notice that, after a single deoxyribonucleotide has been added to the chain, a new 3′-hydroxyl end is left to which the next deoxyribonucleoside-5′-phosphate can be attached.

This in essence is how the new polydeoxyribonucleotide strand grows in the 5′ → 3′ direction. But we say 'in essence' because the replication of DNA is an extremely complicated matter, and the account that we have given so far has failed to mention several important problems. The first problem results from the fact that in a

double-stranded molecule of DNA the two strands are antiparallel
– that is, the 5'-hydroxyl end of one strand is hydrogen bonded, by
complementary base pairing, to the 3'-hydroxyl end of the other
(see p. 54). When a new strand of DNA is being made, by the
reaction we have just described, in the 5' → 3' direction, the enzyme
will be following the old strand along in the 3' → 5' direction. That is
to say, if the new strand has its 3'-hydroxyl end at the bottom of the
page (as in our diagram), so that the enzyme is working from top to
bottom, the old strand that is being copied has its 5'-hydroxyl end at
the bottom of the page. But because of the antiparallel nature of
DNA, the *other* old strand (the original partner of the strand that is
being copied) has its 5'-hydroxyl end at the top and its 3'-hydroxyl
end at the bottom; to copy this second old strand would involve
synthesizing a new strand from bottom to top, since the specificity of
DNA polymerase ensures that it can synthesize new strands only in
the 5' → 3' direction. So it might have seemed that the replication of
double-stranded DNA would necessitate two enzyme molecules
starting at different places and working in opposite directions. But
in fact we have excellent evidence that there is a *single* point of
replication moving along the DNA molecule, and that *both* of the
old strands are copied simultaneously.

The mechanism for achieving replication of both strands is, we
believe, as follows. One of the two new strands is synthesized
continuously by one enzyme molecule, which catalyses the reaction
that we have just described between an incoming deoxyribonucleoside triphosphate and the 3'-hydroxyl end of the new
strand, and ensures that the incoming nucleotide that is selected is
complementary to the opposite nucleotide in the parental strand.
The other new strand is synthesized by another enzyme molecule
according to the same reaction and also ensuring complementarity
of nucleotides. But this second strand is synthesized in small pieces,
so that the enzyme molecules that catalyse synthesis of the two
strands are both working in the same region of the parental DNA
rather than from opposite ends (see Fig. 19.1). These small pieces
have to be stitched together, after their synthesis, by a special
enzyme called a ligase.

A second problem about DNA replication is that the DNA
polymerase that adds nucleotides to the 3'-hydroxyl end of the
growing chain will not start a new DNA chain; it acts only by
extending a *primer* (compare the synthesis of glucose polymers that
we described on p. 165). It turns out that the primer for DNA
synthesis is a short chain of *RNA*, made by a special enzyme that is

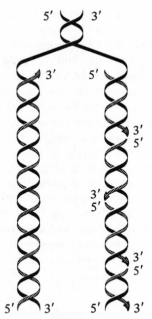

Fig. 19.1. The replication of double-stranded DNA. The overall direction of replication is from bottom to top of the page. The grey lines represent newly synthesized strands. The strand on the left is being synthesized continuously in a 5′ → 3′ direction. The other strand is also synthesized in a 5′ → 3′ direction, but its synthesis is in small pieces. The head of each arrow represents the 3′-hydroxyl end of a newly synthesized molecule.

different from the usual RNA-synthesizing enzyme we shall discuss below. After the DNA polymerase has begun to add deoxyribonucleotides to this primer, the RNA is digested away; this process leaves a gap, which is filled with deoxyribonucleotides by yet another enzyme.

Another problem arises from the fact that DNA is a helical molecule (see p. 54). In order for new strands of DNA to be synthesized, with the incoming deoxyribonucleotides selected by base pairing with the nucleotides of the parental strands, these parental strands must separate. Such separation involves local unwinding, and still further proteins are required to catalyse this unwinding.

There are even more difficulties about DNA replication which we have not mentioned here. We have said enough to indicate that the process is extremely complicated and that it requires many components. But it is important not to lose sight of the basic features of the process, since the outline is simple enough: a new DNA strand is synthesized by polymerization of deoxyribo-nucleoside triphosphates to form a polydeoxyribonucleotide chain, with the incoming nucleotide selected by its ability to pair by hydrogen bonding with a corresponding nucleotide in the parental strand. By this process two double-stranded DNA molecules are formed, which are identical in base sequence to one another and to the parental double-stranded molecule.

We can now turn to the synthesis of RNA, which is analogous in some respects to the synthesis of DNA. An enzyme called RNA polymerase is known which, in the presence of DNA, catalyses the synthesis of ribonucleotide polymers. This enzyme employs nucleoside 5' triphosphates (products of the synthetic pathways we have described on pp. 190–4) and splits out one pyrophosphate for every nucleotide inserted into the polymer; it requires the presence of all four nucleoside triphosphates, and it synthesizes an RNA molecule in the 5' → 3' direction.

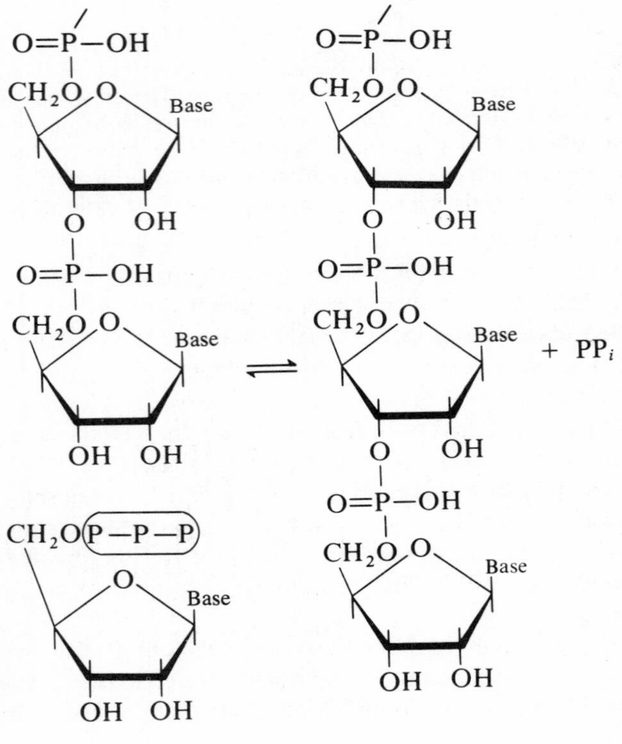

The synthesis of RNA by this enzyme is dependent on the presence of DNA, which acts as a template that the enzyme copies, or 'transcribes'. This transcription involves pairing between the bases of the incoming ribonucleotides that are to form the RNA molecule and the bases of the deoxyribonucleotides in the template DNA. The RNA polymerase, however, transcribes only one strand of DNA, and forms a single-stranded molecule of RNA that is similar in sequence (except that uracil is substituted for thymine) to the *opposite* strand of the DNA. Once again we use bold type to indicate the new nucleotides, this time those that are to form the RNA:

. . . U-U-G-A-C-C-C-C-A-A-G-G-U-A-C
. . . A-A-C-T-G-G-G-G-T-T-C-C-A-T-G
. . . T-T-G-A-C-C-C-C-A-A-G-G-T-A-C.

In this way the enzyme makes available an exact RNA transcript of the DNA. More accurately we should say that the RNA polymerase can make available an exact RNA transcript of any desired length of the DNA. The DNA contains particular sequences of bases at which the RNA polymerase binds to initiate transcription (see also p. 232), and other sequences of bases at which the RNA polymerase stops transcription. The length of DNA between a starting point and a stopping point is transcribed into a single molecule of RNA.

The great importance of this synthesis of RNA is that it enables proteins to be synthesized. In the next chapter we shall discuss in detail the role of RNA in the synthesis of proteins.

20 Synthesis of proteins

We outlined in Chapter 18 some of the evidence that leads to the conclusion that the linear sequence of the bases in DNA determines the linear sequence of amino acids in proteins. But at the end of Chapter 19 we said that it is the RNA transcribed from DNA, rather than the DNA itself, that is actually involved in protein synthesis.

Most of the RNA that cells contain is present in small particles called *ribosomes* (see p. 57). Ribosomes consist of RNA and protein in the rough ratio of 60 : 40; the relative molecular mass of bacterial ribosomes is about 2.6 million, and of ribosomes from higher cells about 4–5 million. One interesting feature of ribosomes is that they seem always to consist of two unequal subunits: in bacteria (the ribosomes of which have been studied in most detail) their relative molecular masses are about 1.8 million and 0.8 million. It is conventional to give ribosomes, and their subunits, names that refer to their rate of sedimentation in a centrifugal field. Thus in bacteria the bigger subunit is called the '50 S particle' and the smaller the '30 S particle', and the whole ribosome is called the '70 S ribosome'. The corresponding names for the ribosomes of higher cells are '60 S particle', '40 S particle' and '80 S ribosome'. The 30 S particle of bacteria contains a single molecule of RNA (see p. 57) and about twenty molecules of protein (probably representing just one copy of each of twenty different proteins). The 50 S particle contains two different molecules of RNA and about thirty-five different molecules of protein.

The reason why we have given this description of ribosomes is that they represent the site of synthesis of proteins; we shall see later just how it is that they function. What we must now consider is whether, as one might expect, it is the RNA of the ribosome that is involved in specifying protein structure. Ribosomal RNA (rRNA), like all RNA, is synthesized by the RNA polymerase which transcribes the base sequence of DNA into a similar base sequence of RNA. Now we have seen that the sequence of bases in DNA ultimately determines the sequence of amino acids in protein. So it would seem reasonable to imagine that the rRNA acts as an intermediate in protein synthesis and that the structure of the protein, determined at one remove by DNA, is *directly* determined by the rRNA.

This idea turns out not to be correct. It is found, for example, that bacteria can change their pattern of protein synthesis without making new rRNA, and there is reason to believe that all ribosomes in a single bacterium contain identical molecules of RNA even though they are synthesizing many different proteins. In other words, the ribosome contains machinery for *assembling* proteins (e.g. for making peptide bonds) but not the instructions for determining the sequence of their amino acids. A ribosome receives such instructions from time to time, and, until they are superseded, will follow them in assembling a protein.

These instructions are provided to the ribosome in the form of another species of RNA, which is called messenger RNA. Messenger RNA (mRNA), like rRNA, is synthesized by the transcribing activity of RNA polymerase (p. 200); but in this case the sequence of nucleotides that is transcribed determines specifically the sequence of amino acids in proteins, whereas the DNA sequence that is transcribed to make rRNA does not correspond to any particular protein. A consequence of this difference is that most of the DNA of an organism, which is occupied with determining the structures of the thousands of different proteins synthesized, is transcribed into mRNA; only a small fraction of the DNA is transcribed into rRNA.

Earlier we used the word 'gene' to describe a length of DNA the sequence of which corresponds to the sequence of a protein. We also showed how RNA polymerase was capable of making an RNA transcript of any desired length of DNA. Putting these two ideas together would enable us to say that RNA polymerase can transcribe a gene to produce the *mRNA* that corresponds to that gene. However, there are two ways in which we must qualify this statement. First, we shall see in Chapter 22 that in bacteria a mRNA may sometimes correspond to several contiguous genes. Secondly, in higher cells the RNA that is transcribed from the DNA is not used as mRNA without further treatment, but seems to have some sections of the molecule removed (in ways that we do not by any means fully understand) before acting as mRNA. Despite these complications, we will not go far wrong in regarding a molecule of mRNA as resulting from transcription of a given gene. By receiving this mRNA ribosomes will be instructed to synthesize the protein that corresponds to that gene. So long as that mRNA remains intact they will synthesize that particular protein, but if after some time the mRNA is degraded (see p. 231) they will be free to accept a fresh mRNA and synthesize a new protein. In a sense, therefore, the ribosome is the slave of whatever mRNA happens to be bound to it at any given time.

So far we have spoken loosely of the mRNA 'instructing' the ribosome how amino acids should be arranged to make a protein. In fact the only specificity that a molecule of mRNA can possess must lie in its sequence of bases, in just the same way as is true of DNA. How then can the sequence of bases in mRNA determine the sequence of amino acids in a protein? In order to answer this question, we must look at the fate of the activated amino acids that we described on p. 178. We left the amino acids in a sort of limbo, attached to AMP; but actually aminoacyl-AMP is tightly held to the enzyme that makes it and is not set free. The amino acids are immediately transferred again from AMP; and the molecules that now accept them belong to a third species of RNA which is called transfer RNA or tRNA (p. 57). This reaction is catalysed by the same set of enzymes as formed the aminoacyl-AMP intermediates; each enzyme is highly specific for a particular amino acid. We can now write the two reactions catalysed by these enzymes (whereas on p. 178 we artificially interrupted the process after the first reaction).

$$RCHNH_2COOH + ATP \rightleftharpoons RCHNH_2CO-AMP + PP_i$$
$$RCHNH_2CO-AMP + tRNA \rightleftharpoons RCHNH_2CO-tRNA + AMP.$$

As pyrophosphate is produced the equilibrium of the total reaction is strongly in favour of the aminoacyl tRNA (see p. 141).

Each molecule of RNA is specific for a particular amino acid, and since the enzymic reaction too is specific for the amino acid it follows that each amino acid is matched to its own tRNA molecule. Once an amino acid has been bound to its specific tRNA, it is the specific structure of the *tRNA* (rather than of the amino acid) that enables the complex to be recognized.

Transfer RNA molecules are exceptionally small among RNA species, containing only seventy to eighty nucleotides (see Fig. 20.1 and Table 6.1). The 3'-hydroxyl end of the molecule (see p. 6) always has the sequence: cytidine-cytidine-adenosine. The amino acid is bound to the 3'-hydroxyl group of the adenosine. Since this binding is common to all tRNA molecules, it cannot represent the specific feature by which each one is recognized.

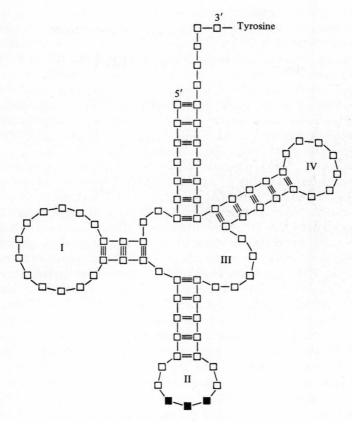

Fig. 20.1. Tyrosyl tRNA.

In fact the recognition of tRNA molecules has two aspects. In the first place the molecule must interact with the specific enzyme that charges it with its amino acid: we do not know in detail how this recognition takes place. Secondly the tRNA molecule must be recognized during the translation of the mRNA by the ribosome, so that the amino acid that it carries can be incorporated in its correct position in the polypeptide. The recognition site for this step is known: it is a sequence of just three bases in the tRNA. In those tRNA molecules whose structure has been worked out it is found that this recognition site (marked in bold type in Fig. 20.1) is always in the second single-stranded region of the chain, counting from the 5′ end.

It might be convenient at this point to recapitulate, since we have now described the three chief reactants in the process that results in the synthesis of proteins. The *ribosome* is the site of synthesis of proteins, and although ribosomes contain RNA, this RNA does not determine the sequence of amino acids that are to be assembled. *Messenger RNA* carries the sequence of bases that dictates the sequence of amino acids. *Transfer RNA* carries the amino acids, and each molecule includes along its length a sequence of three bases which, since it is characteristic of that molecule, signals which amino acid is attached to its 3'-hydroxyl end.

We have seen (pp. 54 and 195) that hydrogen bonding between bases permits a sequence of bases in one polynucleotide chain to recognize a complementary sequence in another polynucleotide chain. We have previously mentioned this interaction in terms of the two strands of DNA, and in terms of pairing between DNA and RNA; but it is equally possible for a sequence of bases in one RNA molecule to recognize by hydrogen bonding a sequence of bases in another RNA molecule. So it should be possible for the sequence of three bases that is characteristic of each tRNA molecule to recognize a complementary sequence of three bases in mRNA. This idea suggests that sets of three bases in each molecule of mRNA might be used to specify which amino acids should be incorporated into the corresponding protein. Since there are four different bases in RNA, there are $4 \times 4 \times 4$ ($= 64$) possible combinations of bases, and this is more than enough to account for all the amino acids that need to be specified.

It is now certain that this is the method by which mRNA dictates which amino acids are to be assembled into the polypeptide chain. Each of the amino acids corresponds to a sequence of three bases in mRNA, and each of the possible three-base sequences is known to specify a particular amino acid (except for three of the sixty-four, which are used to signal the end of the polypeptide chain). Each of these three-base sequences, which specifies or 'codes for' an amino acid, is called a *codon*; the complementary sequence of three bases in each tRNA, which is characteristic of the molecule of tRNA and signals the nature of the amino acid that it is carrying, is called the anticodon. Table 20.1 gives the assignment of codons to amino acids.

So far we have said little about the role of the ribosome. As we mentioned earlier, ribosomes are large and complicated organelles, and they are being intensively studied at present. The results of these studies have shown that ribosomes have many functions. One

Table 20.1

UUU ⎫ UUC ⎬	Phenylalanine	UCU ⎫ UCC ⎪	
UUA ⎫ UUG ⎬	Leucine	UCA ⎬ UCG ⎭	Serine
CUU ⎫ CUC ⎪ CUA ⎬ CUG ⎭	Leucine	CCU ⎫ CCC ⎪ CCA ⎬ CCG ⎭	Proline
AUU ⎫ AUC ⎬ AUA ⎭	Isoleucine	ACU ⎫ ACC ⎪ ACA ⎬	Threonine
AUG	Methionine	ACG ⎭	
GUU ⎫ GUC ⎪ GUA ⎬ GUG ⎭	Valine	GCU ⎫ GCC ⎪ GCA ⎬ GCG ⎭	Alanine
UAU ⎫ UAC ⎬	Tyrosine	UGU ⎫ UGC ⎬	Cysteine
UAA	*	UGA	*
UAG	*	UGG	Tryptophan
CAU ⎫ CAC ⎬	Histidine	CGU ⎫ CGC ⎪	
CAA ⎫ CAG ⎬	Glutamine	CGA ⎬ CGG ⎭	Arginine
AAU ⎫ AAC ⎬	Asparagine	AGU ⎫ AGC ⎬	Serine
AAA ⎫ AAG ⎬	Lysine	AGA ⎫ AGG ⎬	Arginine
GAU ⎫ GAC ⎬	Aspartic acid	GGU ⎫ GGC ⎪	
GAA ⎫ GAG ⎬	Glutamic acid	GGA ⎬ GGG ⎭	Glycine

* These codons signify termination of the polypeptide.

function is to bring the mRNA into precise spatial juxtaposition with a succession of aminoacyl tRNA molecules. A second function is to catalyse, at two stages in the incorporation of each amino acid, the hydrolysis of GTP to GDP and inorganic phosphate. A third function, as we shall see, is to hold the growing peptide chain.

We can describe the mechanism of assembly of amino acids into protein by using as an example the translation by a bacterial

ribosome of the small fragment of mRNA whose synthesis we
described in the last chapter. This mRNA has the sequence
... UUGACCCCAAGGUAC.... We have already seen that a
sequence of three bases in a mRNA molecule specifies one amino
acid, so that we can represent the fragmentary mRNA as
... UUG ACC CCA AGG UAC .... By reference to Table 20.1
we can see that the corresponding peptide will be ... leucyl-
threonyl-prolyl-arginyl-tyrosine .... It is easiest to describe the
process of *elongation* of the peptide chain, in other words to imagine
the process in full swing, and we shall assume that the first codon
(UUG) of our mRNA has just been translated – i.e. that the leucine
has just been inserted into the polypeptide – so that the codon ACC
is now waiting to be translated.

We can represent this situation by the diagram given as Fig. 20.2.
The mRNA is bound to the 30 S subunit of the ribosome; the part of
the mRNA that is to the left of the ribosome has already been
translated, and the part that is to the right of the ribosome is going to

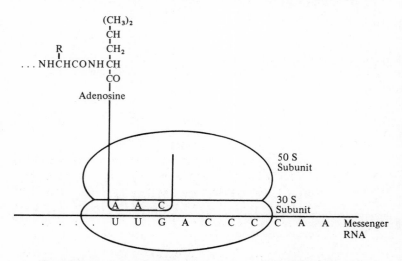

Fig. 20.2. Protein synthesis, stage 1. For explanation, see text.

be translated. The ribosome is holding a molecule of peptidyl
tRNA; this represents the incomplete peptide that has already been
synthesized, terminating in leucine and esterified to the leucine-
specific RNA. The last point will become clearer shortly; at this
point what we wish to stress is that the peptide has grown from its

amino end so that it can be represented as:

$$
\overset{R_1}{|} \quad \overset{R_2}{|} \quad \overset{R_3}{|} \quad \overset{R_4}{|}
$$
.....NHCHCONHCHCONHCHCONHCHCO—tRNA.

The ribosome will now accept another aminoacyl tRNA – that which is specified by the codon ACC (threonyl tRNA) – and the resulting situation will be that represented in Fig. 20.3.

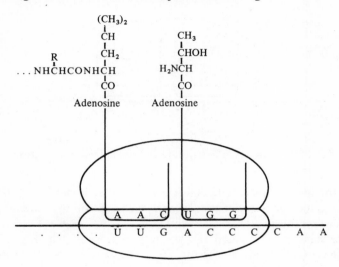

Fig. 20.3. Protein synthesis, stage 2. For explanation, see text.

Immediately after binding the aminoacyl tRNA, the ribosome catalyses the hydrolysis of a molecule of GTP, and this hydrolysis appears to change the conformation of the ribosome–aminoacyl tRNA complex, so that the ribosome is now in a position to catalyse the next reaction. This reaction is the formation of a peptide bond between the carboxyl group of leucine and the amino group of threonine. In other words the polypeptide terminating in leucine is transferred to threonine, so that the leucine-specific tRNA through which the polypeptide was held is no longer esterified (Fig. 20.4).

The ribosome now moves one codon with respect to the mRNA, meanwhile holding the lengthened peptide, which terminates in threonyl tRNA, opposite the threonine codon of the mRNA, so that the relative position of the tRNA and mRNA remains unchanged. In this *translocation* another molecule of GTP is hydrolysed, and

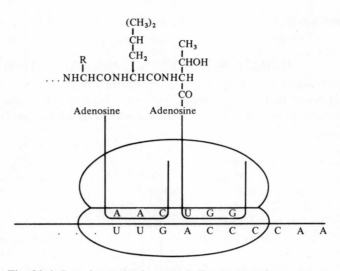

Fig. 20.4. Protein synthesis, stage 3. For explanation, see text.

also the leucine-specific tRNA is ejected; later it can be esterified again to leucine in order to take part once more in the reactions of protein synthesis.

We have now reached the position represented in Fig. 20.5, which is exactly analogous to Fig. 20.2. The next step will be binding

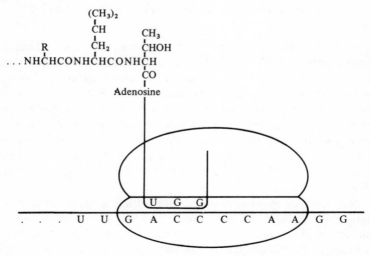

Fig. 20.5. Protein synthesis, stage 4. For explanation, see text.

of prolyl tRNA specified by the codon CCA. By repetitions of the series of reactions that we have described, proline, arginine and tyrosine will be successively inserted into the peptide. Notice that the incorporation of each amino acid necessitates the hydrolysis of two molecules of GTP in addition to the loss of two high-energy bonds in the formation of aminoacyl tRNA (p. 178).

We may now consider the mechanism for *termination* of synthesis of a polypeptide. After it has inserted the final (C-terminal) amino acid of a polypeptide, the ribosome encounters on the mRNA one of three chain-terminating codons – UAG, UAA and UGA – that do not correspond to any aminoacyl tRNA. The special chain-terminating codon is recognized by a protein 'release factor', which causes hydrolysis of the bond between the completed polypeptide and the tRNA through which it is held to the ribosome. The completed polypeptide is thus released, and the complex consisting of the ribosome, the final tRNA and the mRNA is then dissociated.

The description that we gave above for elongating a peptide chain involved the presence of a peptidyl tRNA on the ribosome throughout the cycle of events illustrated in Figs. 20.2–20.5. Our description therefore does not account for the *initiation* of synthesis of a peptide, and in fact it turns out that initiation is quite different from elongation. It involves not the 70 S ribosome but instead the separate 30 S and 50 S subunits. It also involves a special initiating

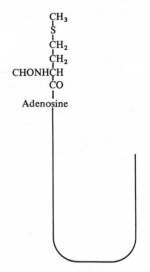

Fig. 20.6. *N*-Formylmethionyl tRNA.

tRNA called *N*-formylmethionyl tRNA, which is more analogous to a peptidyl tRNA than to an aminoacyl tRNA (see Fig. 20.6). The first step in the process is the binding of a 30 S particle to the mRNA at an initiating codon, AUG. In response to this codon, *N*-formyl-methionyl tRNA joins the complex, and then a 50 S ribosomal subunit too is bound. We have now reached a position similar to that illustrated in Fig. 20.2, the only difference being that, instead of the peptidyl tRNA shown in that Figure, the ribosome (assembled from its two subunits) has attached to it *N*-formylmethionyl tRNA. Next another aminoacyl tRNA is bound, exactly as shown in Fig. 20.3, and from that point the normal process of elongation can occur.

The polypeptides, when they are released from the ribosome after synthesis is complete, may still have the *N*-formyl group attached to them. The formyl group, or the whole of the *N*-formylmethionine, is removed by hydrolysis with a special enzyme.

As each ribosome, after initiating synthesis, moves along the mRNA to elongate the polypeptide, it leaves free the initiation site on the mRNA at which it became attached. Another 30 S ribosomal subunit can then bind to this point, followed by a molecule of *N*-formylmethionyl tRNA and a 50 S subunit, and this second 70 S ribosome too will start to move along the mRNA, translating as it goes. In this way a structure called a polyribosome is built up (see Fig. 20.7). Protein synthesis thus generally involves a number of ribosomes which, although they are at different stages in the process, are engaged in simultaneous translation.

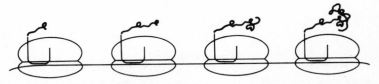

Fig. 20.7. A polyribosome. Ribosomes move, with respect to the mRNA, from left to right, and the polypeptide chain that they are synthesizing is thus gradually lengthened.

Section IV
Compartmentation and regulation

21 The cell and membrane transport

Although all the reactions that we have described in Sections II and III of this book take place in cells, we have hardly discussed at all how these reactions are *localized* within the cell or how different parts of the cell are specialized to serve different functions. In this chapter we shall give an outline of these features of cellular metabolism.

Cytology – the study of the structure of cells – has become in the last few decades an enormous subject, related to but to some extent independent of biochemistry. In this chapter we can do no more than refer to a few of the principles that have been developed, concentrating in particular on the way in which metabolic pathways are distributed within the cell. We shall see that there is considerable specialization *within* the cell: the enzymes of the different pathways are not distributed evenly throughout the whole volume of the cell, but rather are concentrated in one or other place. We can think of this localization as having a regulatory function. By segregating the enzymes of a particular pathway from those of another pathway, and ensuring that metabolites cannot move freely from one to the other, the cell can control the rate at which compounds are metabolized by particular routes. This means of regulation is only one of those that organisms have at their disposal: in the next two chapters we shall consider two others.

There are very many different types of cell among the range of living organisms. The first distinction to be made is between *procaryotic* and *eucaryotic* cells. The important characteristic of *procaryotic* cells, which belong to such organisms as bacteria, is that they contain no internal membranes – that is no membranes within the cell itself – although, like eucaryotic cells, they have a limiting membrane that surrounds the cell. Eucaryotic cells have internal membranes which separate off individual parts of the cell from one another. Some of these parts of the cell have characteristic shapes and, as we shall see, characteristic complements of enzymes; thus they form well-defined membrane-bound bodies which are called *organelles*. Eucaryotic cells occur in all true plants and animals, and also to many lower organisms such as fungi and some algae.

We shall not discuss procaryotic cells further. Among eucaryotic cells we shall say only two brief things about plant cells – namely that they are characterized by having thick cell walls made of cellulose (see p. 67), and that apart from the fact that some of them have chloroplasts (p. 99) most of their organelles are quite similar in function and enzymic complement to those of animal cells.

Having in a few words greatly narrowed the scope of this chapter, we shall now narrow it still further by concentrating on one particular type of animal cell. In the body of the animal, there are many different tissues that serve different functions. Many of these are highly specialized, and the cells of which they are composed are directed towards just one function (for example the synthesis of a hormone – see p. 223). Other cells, however, are far less specialized: they carry out many biochemical activities including most of the pathways that have been described in this book. Thus in discussing the localization of the enzymes involved in these pathways we can most conveniently use for our description one of these relatively unspecialized cells. The best one to take – best in the sense of having an abundance of enzymic activities – is the *hepatocyte*, which is the type of cell of which the liver is chiefly composed, the liver being the organ that is responsible for much of the degradative and synthetic activity of the body.

Fig. 21.1 represents a thin section through a hepatocyte, as viewed with the electron microscope, and it shows an abundance of fine structure. The identification of some parts of the cell is given on the Figure. Although we might be tempted to describe the cell as composed of an homogeneous liquid in which the organelles are suspended, this description would be inaccurate. The liquid – often called the soluble fraction of the cytoplasm or *cytosol* – is not a simple homogeneous solution, but is itself known to contain aggregates of molecules even though these cannot usually be seen with the electron microscope. In addition, there are several organelles of the cell that can be clearly distinguished from the cytosol: the *nucleus*, the *mitochondria*, the *endoplasmic reticulum* and the *lysosomes*. We shall say something about each of these in turn.

The *nucleus* is that organelle in which almost all of the genetic information of the cell is present. A small amount of DNA is found in mitochondria (and in chloroplasts of plant cells), but it is the nucleus that is responsible for carrying all but a very small fraction of the genetic complement of the animal cell. In the hepatocyte the nucleus accounts for about 10% of the mass of the cell (in rapidly

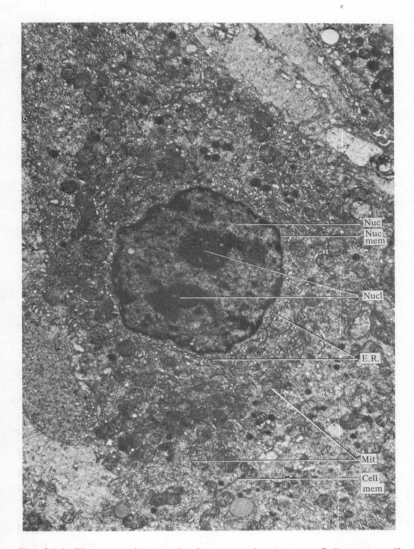

Fig. 21.1. Electron micrograph of a mouse hepatocyte. Cell mem = cell membrane; E.R. = rough endoplasmic reticulum; Mit = mitochondria; Nuc = nucleus; Nuc mem = nuclear membrane; Nucl = nucleoli. (The photograph was kindly supplied by Dr J. C. F. Poole, Sir William Dunn School of Pathology, University of Oxford.)

dividing cells this figure can approach 50%); although DNA comprises only some 15% of the nucleus, the distinguishing feature of the nucleus is that it contains the cell's genetic information. In addition to DNA, nuclei contain various proteins. Some of these are the basic proteins called histones, which (as we saw on p. 00) form complexes with DNA to make up chromatin. In addition there are many other proteins in the nucleus, of which some may have structural roles while others (such as the DNA and RNA polymerases we discussed in Chapter 19) have enzymic functions.

We saw in Chapter 19 that a vital function of DNA was to act as a template for the synthesis of RNA. Of the different kinds of RNA made in the cell much the most abundant is ribosomal RNA (rRNA), and therefore one of the major functions of the nucleus is to manufacture rRNA. There is good evidence that this manufacture occurs in a specific region of the nucleus called the *nucleolus*, which is readily visible in electron micrographs (see Fig. 21.1) as a dense area. Sometimes there are two nucleoli, or occasionally more. Nucleoli are especially prominent in cells that are synthesizing proteins particularly rapidly, which, considering the role of RNA in the synthesis of protein, is what one would expect.

The nucleus is separated from the cytoplasm by a nuclear envelope, which is composed of a double membrane system that may be part of the endoplasmic reticulum (see below). At many places the nuclear envelope is interrupted by pores; these appear to allow communication between the nucleus and the cytoplasm, but it is probable that the flow of material in and out of the nucleus is controlled in some way and that there is not simply free diffusion between nucleus and cytoplasm.

A single hepatocyte contains about 1000 *mitochondria*, and as we saw in Chapter 8 these too are bounded by a double membrane. The outer membrane is smooth and regular, but the inner membrane is extensively folded, and the folds, which are known as cristae, project inwards and extend through much of the mitochondrial space. On the inner face of the inner membrane there are many small spheres, which are attached to the membrane by stalks. The ATP-ase–proton translocase system that we mentioned in Chapter 8 is associated with these structures.

Mitochondria are composed of protein (about 70%) and lipid (about 30%); there are also small amounts of DNA and RNA present. Many important enzymic activities are located in the mitochondria. The matrix (the part of the mitochondrion within the inner membrane) contains enzymes of the Krebs cycle (Chapter 11) and the enzymes for oxidizing fatty-acyl coenzyme A (Chapter 12). The inner membrane has the components of the respiratory chain and the enzymes necessary for oxidative phosphorylation (Chapter 8); in fact it is possible to isolate small, membrane-bound particles by disrupting the inner mitochondrial membrane; these particles, which are still closed systems, bounded by a membrane, can be made to carry out oxidative phosphorylation coupled to electron transport.

Thus mitochondria contain the systems necessary for synthesizing ATP by oxidation of the intermediates of carbohydrate and fat degradation. We have seen in Chapters 11 and 12 that these are by far the most important sources of ATP for aerobic organisms; the mitochondrion well deserves its popular nickname 'the power-house of the cell'. The fact that these activities are localized within a comparatively small organelle clearly makes for rapid and efficient oxidative phosphorylation at the expense of the metabolic inter-mediates of carbohydrate and fat degradation. At the same time it gives rise to problems of intracellular transport. How do the inter-mediates gain access to the inside of the mitochondria? How does ATP, made in the mitochondria, get out – for example to drive the synthetic reactions such as protein synthesis that are taking place at ribosomes outside the mitochondria? These questions will be considered later in this chapter.

Running through the cytoplasm is a system of tubules and sacs enclosed by membranes, with the result that a network of membrane-bounded spaces penetrates the whole cell. This network is called the *endoplasmic reticulum* (E.R.). Sometimes the membranes of the E.R. are studded on their outer surface with ribosomes (the reticulum is then called *rough E.R.*) and sometimes the membranes have no ribosomes (*smooth E.R.*).

The endoplasmic reticulum contains certain enzymic activities, for example the enzymes for synthesizing steroids and for esterify-ing fatty-acyl coenzyme A to form triglycerides. A further important function of the E.R. is to collect and process materials that are destined for export from the cell. For example, proteins that are to be released from the cell are synthesized by the ribosomes that line the rough E.R., and are then collected within the spaces of the E.R. and concentrated into vacuoles which finally disgorge their contents at the surface of the cell. Sometimes the material to be secreted is modified during its stay in the E.R., or in another organelle called the *Golgi apparatus.* (This is somewhat similar in structure to the E.R.; it consists of closely spaced sacs bounded by membranes.) We can exemplify such modifications by mentioning the secretory proteins known as glycoproteins, which are proteins to which carbohydrate is attached (p. 68); this addition of carbohydrate occurs in the Golgi apparatus.

Lysosomes are small particles (about 0.5 μm in diameter), bounded by membranes. Their chief feature is that they contain hydrolytic enzymes. These enzymes are capable of breaking down the large molecules found in living tissues: thus they include enzymes that hydrolyse proteins, nucleic acids, polysaccharides, lipids, etc. The lysosomes appear to arise from the Golgi apparatus; it is likely that the hydrolytic enzymes are made by the ribosomes of the E.R. and packaged at the Golgi apparatus into small membrane-bound bodies which become the lysosomes.

One function of lysosomes is to digest macromolecules arising from outside the cell. Sometimes molecules become adsorbed to the external membrane of the cell, and the membrane then folds round them, enclosing a droplet of the extracellular medium in a vacuole; this process is called pinocytosis. The vacuole may then fuse with a lysosome, and its contents will become digested by the hydrolytic enzymes. In addition lysosomes sometimes digest macromolecules arising within the cell, and it is thought that this process may contribute to the turnover of cellular constituents that is a marked feature of all living organisms; however, we do not know how the cell chooses the material that is to be digested. Again, after a cell has died, the contents of lysosomes are apparently released and act in hydrolysing its macromolecules to form simpler molecules which can be used again as building blocks.

The *cytosol*, which as we saw above is the phase of the cell in which the organelles are suspended, also has characteristic enzymic activities. These include the enzymes of glycolysis (as opposed to those of the Krebs cycle which, as we just mentioned, are associated with the mitochondria). Another activity found in the cytosol is the synthesis of fatty acids, and many of the enzymes of amino-acid synthesis and degradation. We should not think of the cytosol as resembling a thin, watery soup; with its large number of intermediates, and enzymes, in solution, the total concentration of solutes in the cytosol is very high.

Transport across membranes

Having emphasized the importance of the different compartments of the cell we may now examine the different modes of transport of materials between one compartment and another, and between the inside and the outside of the cell.

Anyone who has studied the chemistry of non-living materials will be familiar with the phenomenon of *semipermeability*. This property is shown by certain substances which form sheets with microscopic pores. Molecules that are small enough can pass through the pores while larger molecules are prevented from doing so.

Biological membranes, too, show this behaviour, although only a few of the many substances that are transported across them travel by this means. When the process does occur, it is called *free diffusion*. An example is the movement of small, uncharged hydrophilic molecules (such as glycerol and urea) across the outer membranes of cells from a region of higher concentration into one of lower concentration. The movement, and the dependence of rate of transport on molecular size, can be completely explained by assuming the membrane to be permeated by pores approximately 0.8 nm in diameter.

A second type of free diffusion depends not on the existence of pores, but on the degree of hydrophobicity of the substance that has to pass. A molecule that is sufficiently apolar will merge with the membrane and can leave it on the other side if the concentration there is lower. For instance, diacetyl glycerol passes across cell membranes nearly ten times faster than monoacetyl glycerol, which in turn passes more than ten times faster than glycerol itself. The decrease in rate of passage here correlates quite well with the progressive decrease in hydrophobicity. If we decrease hydrophobicity still further (e.g. by phosphorylating an –OH group) the resulting compound will hardly diffuse through at all.

Certain substances pass through the membranes much more rapidly than either their size or lipid solubility would lead one to expect. These substances enter by what are known as *mediated transport* systems. Transport of this kind differs from free diffusion in two important respects, both of which give us a clue as to the molecular nature of the events taking place. The first of these concerns the effect on the rate of transport of an increase in concentration of the substance to be transported. In free diffusion, increases in concentration lead to a directly proportional increase in rate (e.g. twice the concentration, twice the rate) and within reasonable limits this simple relationship is maintained no matter how great the increase in concentration. By contrast, although mediated transport systems also tend to show an increase in rate in response to increases in concentration, the relationship is not so simple. Only when the initial concentration is extremely low will, say, a doubling in concentration lead to a doubling in rate. Each successive rise in concentration has less and less effect on the rate of mediated transport, until a concentration is eventually reached above which any further rise will bring about no increase whatever in rate of transport. The system is then said to be *saturated*. The second difference between free diffusion and mediated transport is that the latter shows a much higher degree of *discrimination* between different candidates for transport. Instead of using size and lipid solubility as the sole criteria, mediated systems have a high degree of specificity. They can show preference for, say, one optical isomer of a substance over another (e.g. for D rather than L sugars).

Both the specificity and the tendency to saturation of mediated systems lead irresistibly to the conclusion that some association is occurring between the added substance and a limited number of receptor structures. Saturation occurs when all the receptors are fully employed all the time, and specificity is explained by the receptors being able to discriminate between different molecules by specific binding interactions of precisely the kind that we discussed in Section I of this book. The specificity is particularly reminiscent of that of the proteins: it even includes the existence of competitive inhibitors that are structurally related to the molecules that are transported (cf. Fig. 5.4). We may conclude that we have discovered the function of at least some of the protein molecules in the membrane, which we mentioned on p. 85, and it has in fact become possible to isolate and characterize some of the transport proteins.

Mediated transport can take place *passively*, that is *down* a concentration gradient, much as free diffusion does, but, remarkably, it can for some substances take place *actively*, that is *up* a concentration gradient. By this we mean that active transport can move molecules into a region in which they are at a higher concentration than in the region that they left. Before we discuss this extraordinary phenomenon, we will look at an example of passive transport in a little more detail.

Much of the experimental work on transport across membranes has been done with sugars. The transport of sugars can be a passive process, so that it occurs only so long as the concentration on the input side is greater than the concentration on the output side. It is the concentration difference that provides the free-energy gradient that permits the transport to occur. In such processes, after the concentrations on the two sides have become equal, the rates of passage are equal and opposite and hence no *net* transport is observed (cf. the approach to chemical equilibrium, p. 94). The choice between the sugars that are allowed to use this means of passing across the membrane and those that are not allowed to use it, or do so with difficulty, has been shown to depend on whether or not the sugar possesses a particular arrangement of –OH groups. This kind of preference is precisely that which one would expect of a protein specificity site. We can imagine hydrogen bonds running from amino-acid side chains to the –OH groups, and possibly some hydrophobic contact with those parts of the under-surface of the sugar that are free of –OH groups. This specificity has a use beyond simply deciding which sugars shall enter the cell; it can also prevent their leaving again. Thus glucose, the major form in which carbo-hydrate enters the cell to be metabolized, is phosphorylated upon arrival and in this form it no longer has the specific structure necessary to pass back across the membrane. This is the converse of the situation that we mentioned a short while ago in which materials destined for *release* from the cell via vacuoles were marked ·'for export' by the attachment of carbohydrate chains (see p. 220).

At their simplest these passive systems are concerned with the passage of only one substance, or members of a related group of substances. Such systems are called *uniports*. Other systems, however, will not function unless two different substances are transported *simultaneously*. It seems that both substances have to bind to the receptor structure before it will act as a shuttle. If the two substances travel together in the same direction the system is called a *symport*, while if they travel in opposite directions the system is called a *antiport* (Fig. 21.2). We shall meet some examples of symport systems a little later on. An example of an antiport system is the ATP–ADP carrier of mitochondria, which transports one molecule of ADP into the mitochondrion as it removes a molecule of ATP from inside. (The utility of such a carrier, in view of the role of the mitochondrion in ATP formation, will be obvious.) The carrier cannot function unless both molecules are available, each on opposite sides of the membrane.

The thermodynamic driving force that we have discussed hitherto for the passage of a substance has been a gradient of concentration, and thus a gradient of free energy down which it can fall. If, on the other hand, we start with concentration differences across the membrane that produce a tendency to move *against* the direction in which net transport is required (i.e. an unfavourable ΔG), how are we to proceed? The answer is the same as it was with the

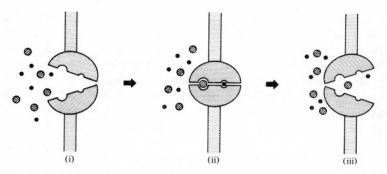

Fig. 21.2(a). A hypothetical view of a symport system. (i): The carrier system is as yet unoccupied, and is open towards the side of the membrane on which there are high concentrations of the substances to be transported. (ii): One molecule each of the two substances appropriate to the particular symport system have collided with their specific receptors and have been bound to them. When both substances have bound the carrier is able to undergo a structural change and in (iii) has opened toward the side on which there is a low concentration of the substances. Because the concentrations are low, the substances will tend to dissociate from their receptors and the carrier will be free to engage in another cycle.

thermodynamically unfavourable chemical reactions in Chapter 7 – by *coupling*. Transport of this kind is, as we have said, called 'active'. The free-energy change necessary to drive active transport can be that associated with the changes in concentration of another substance. Alternatively, as is so often done in biology to overcome energetic problems, coupling can be to the free energy of hydrolysis of high-energy compounds (notably ATP).

The best understood instances in which the fall in concentration of one substance is coupled to the rise in concentration of another involve symport systems. We can imagine that the falling substance drives a water wheel and that the rising substance joins it on the paddles for a free ride. For instance, in certain cells of the vertebrate gut, glucose, and several amino acids, are transported against a concentration gradient, by systems in which the driving energy is provided by the downhill passage, in the same direction, of Na^+. (This system is different from the passive uniport for sugars that we mentioned on p. 223.) The best understood instance in which the consumption of ATP is coupled to a forced rise in concentration is an antiport system known as the *sodium pump*, which transports Na^+ in one direction and K^+ in the other. This system, which is found in most animal cells, accumulates K^+ within the cells and

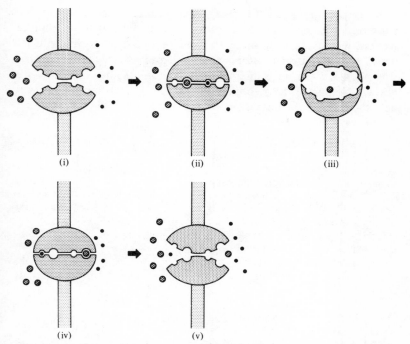

Fig. 21.2(b). A hypothetical view of an antiport system. (i) and (ii): The molecules of the two types approach the carrier and bind to specific receptors. As in Fig. 21.2(a), the carrier system undergoes a conformational change once both substances have bound, but in this case it permits them to move to the sides opposite to those from which they arrived (iii and iv). In (v) the carrier system has opened in response to this movement of the two substances, which then diffuse away into regions in which they are at a low concentration. The carrier system is once again ready to function.

expels Na^+. Since the concentration of K^+ is higher within cells, while the concentration of Na^+ is higher outside, both ions are moved against a concentration gradient. The sodium pump is an extremely important part of the cell's economy; it normally accounts for at least 25% of total energy expenditure. A description of the uses to which these ion gradients are put, and thus the justification of the expenditure of such a proportion of the cell's energy resources on maintaining them, is beyond the scope of this book. We may note, however, that the symport systems that we described as being responsible for the active transport of glucose and amino acids rely on the Na^+ gradient maintained by the pump.

Beyond emphasizing the likely involvement of specific binding sites on proteins, and some rather vague references to paddle wheels, we have said nothing about the molecular mechanisms by which active transport systems operate. As in several earlier parts of this book we are once again forced to confess that the requisite knowledge has yet to be obtained. It seems likely that the key will lie, as it did with the passive mediated transport systems mentioned

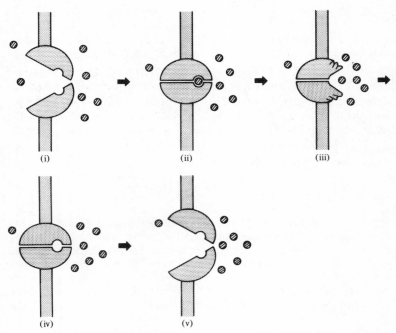

Fig. 21.3. A hypothetical view of an active transport system. For simplicity a uniport is shown. (i) and (ii): The substance to be transported binds to its receptor, having approached from the side on which it is at the lower concentration. (In Fig. 21.2 it was supposed that the strength of the binding to the receptor was weak enough for there to have been a strong tendency for the molecule to dissociate into a region of low concentration, but this does not have to be a universal rule. Often, as in the case suggested here, a substance may have a sufficient affinity for its specific receptor to bind even from a solution in which it is at low concentration.) (iii): The carrier system undergoes a conformational change which distorts the receptor and lessens its binding affinity for the molecule, which is permitted to leave only into the region of high concentration. The energy-requiring nature of this change is suggested by the strain indicated in (iii). Once the strain in the carrier is relieved the system can prepare to accept another molecule (iv and v).

above, in conformational changes in the carrier proteins, particularly in conformational changes that affect quaternary structure. Fig. 21.3 takes the ideas of Fig. 21.2 a step further to indicate in outline one way in which energy-dependent changes in conformation could control the movement of solute substances across the membrane. The type of channel shown in these two Figures is easily formed when a quaternary structure is built up from roughly cylindrical subunits (cf. Fig. 3.7). The energy-requiring conformational changes in Fig. 21.3 would presumably be brought about by the binding of solutes (when coupling is to concentration changes) or by covalent interaction with ATP (when coupling is to the consumption of high-energy compounds). We saw on p. 42 an instance of a major change in quaternary structure in response to binding of a solute and on p. 34 an instance of a major change in quaternary structure during the consumption of ATP. It seems very likely that the principles that we learned in Section I of this book will prove a reliable guide to the interpretation of the results of the research that is now being so widely and vigorously pursued into this vital property of living matter.

22 Control of enzyme synthesis

In Section II of this book we outlined a number of metabolic reactions that cells are capable of carrying out, and you will no doubt have realized that this almost bewildering complexity demands careful organization and control. Similarly in Chapters 19 and 20 we have explained the way in which the information that resides in the DNA can be expressed in terms of protein structure, and once again you may have found yourself wondering how this complicated machinery is regulated.

One device that helps to regulate metabolism is the segregation of different metabolic pathways into different organelles, which we discussed in Chapter 21, but there are two further mechanisms that serve regulatory functions, and this chapter and the next will be devoted to a discussion of them. In this chapter we shall consider the means by which the rate of enzyme *synthesis* is controlled; in the next we shall consider the means by which enzymic *activity* is controlled.

Although it is essential for the genetic material of a cell to contain all the information for specifying the structure of any protein that the cell is capable of making, at no time will the cell find a use for every one of the proteins that it can potentially make. We can illustrate this idea most easily by reference to bacteria. Wild-type *Escherichia coli* cells can synthesize all the amino acids from extremely simple starting materials – glucose, ammonia and a few inorganic salts. In order to synthesize (for example) histidine, they are obliged to make ten enzymes that are specific to the pathway of histidine biosynthesis (see p. 186). If, however, they find themselves in a medium in which histidine is present they can take up histidine from the medium, and in these circumstances they will no longer need to make the ten specific enzymes. A slightly different illustration is provided by the system for the utilization of lactose in *Escherichia coli*. Cells of this species are potentially capable of breaking down the disaccharide lactose, and they have the genetic information necessary to specify the structure of an enzyme (β-galactosidase) that hydrolyses

228

lactose to the two hexoses, glucose and galactose, both of which can be metabolized by the Embden–Meyerhof pathway. However, if they are growing in glucose rather than lactose they will not need to make β-galactosidase since they have a direct source of hexose.

We have just remarked that in the presence of histidine there is no need for *E. coli* to make the enzymes of the histidine biosynthetic pathway, and that in the presence of glucose and the absence of lactose there is no need for them to make β-galactosidase. To say that there is no need, however, is to understate the case. We have seen in Chapters 19 and 20 that the processes of messenger RNA synthesis and protein synthesis are complicated and energetically costly. To make proteins that are not needed, therefore, is not merely neutral in effect but disadvantageous. So there is good reason for cells to have a mechanism that actually *ensures* that superfluous proteins are not made.

The control of protein synthesis is a large subject, and we know a good deal about only a small fraction of it. We have some very clear information about the control of protein synthesis in bacteria, and much of this chapter will be devoted to describing the means by which this control is exerted. We know much less about mechanisms of control of enzyme synthesis in higher cells, and we shall see that in fact the problems that higher organisms have to face are quite different from those confronting bacteria.

Before we describe in a little detail the system for regulating protein synthesis of which we have the most detailed understanding, it will be useful to review some of the possible means of control that organisms might, in principle, use. They can be divided into two classes – control at the level of mRNA synthesis and control at the level of ribosomal function.

In our description (Chapter 20) of the mechanism of protein synthesis we tacitly assumed that the ribosome would translate without discrimination any mRNA that was present in the cell. If this assumption is correct, then the rate of synthesis of any particular protein will depend only on the concentration of its corresponding mRNA; the concentration of any mRNA will in turn depend on the rate of its synthesis and the rate of its degradation. On the other hand, it is possible that even if several kinds of mRNA are present in a cell at the same concentration, one might be translated by the ribosome at a greater rate than the others – in other words, that some control is exerted during the functioning of the ribosome.

So we have three potential points of control: the rate of synthesis of the mRNA corresponding to a particular protein, the rate of degradation of this mRNA, and the rate of translation of the mRNA by the ribosome. There is much more experimental evidence that relates to the first of these possibilities than to the other two, and in the bacterial system that we shall describe it is the control of mRNA synthesis that is more or less exclusively involved in regulating the synthesis of specific proteins.

Bacterial systems

It is no accident that the organism in which control of enzyme synthesis has been best studied is *E. coli*, since that is the species whose genetic structure is exceptionally well understood. Elucidation of the lactose system of *E. coli*, which we shall now describe, has been the result of an extremely fruitful and brilliant collaboration between biochemistry and genetics.

When wild-type cells of *E. coli* are grown in glucose medium, they make extremely small quantities of β-galactosidase (the enzyme that catalyses the hydrolysis of lactose to glucose and galactose). When lactose replaces the glucose, the rate of synthesis of β-galactosidase increases by many hundred times. This phenomenon (the increase in rate of synthesis of an enzyme) is called induction, and a substance that, when present in the medium, causes induction is called an inducer. Although lactose is a powerful inducer of β-galactosidase it is not the most powerful known; various synthetic compounds, all of which are chemical analogues of lactose, are inducers, and some are more powerful.

β-Galactosidase is essential to enable cells to grow on lactose, and if one selects mutants of *E. coli* that cannot grow on lactose one finds that most of them are unable to synthesize β-galactosidase. Some of them, however, are lacking not β-galactosidase, but another protein which is called galactoside permease and which is essential for the uptake of lactose from the medium into the cells. Experiments with inducers have shown that β-galactosidase and galactoside permease are, in wild-type cells, always induced *together*. In the absence of inducer both are made at a very low rate; in the presence of inducer both are normally made at a very high rate; but in the presence of a low concentration of inducer, sufficient to cause only partial induction, both are made at an intermediate rate. The synthesis of the two proteins must therefore be controlled together.

It is possible to find another kind of mutant that is able to synthesize both β-galactosidase and the permease but is unable to *control* this synthesis. As a result both proteins are made by these mutant cells even in the absence of inducer; this phenomenon is called constitutive synthesis and the strains are called constitutive mutants.

Now by genetic techniques it is possible to map the genes for β-galactosidase and for galactoside permease, and these are found to be contiguous. It is possible, too, to study genetically the mutants that synthesize the two proteins constitutively; each of these is found to have a mutation in a gene that is located near to, but not contiguous with, the genes for β-galactosidase and for galactoside permease. Moreover, it is possible to prove (also by genetic techniques) that wild-type strains are capable of synthesizing a substance which constitutive strains cannot synthesize, i.e. that a strain is constitutive *because it lacks* this product. It follows that in wild-type strains in the absence of inducer the synthesis of β-galactosidase and galactoside permease is normally prevented or *repressed*; the substance that acts in this way is known as the repressor. The repressor is found to be a protein, which is absent from constitutive strains.

So far we have described three proteins, β-galactosidase, galactoside permease and the repressor, and their corresponding genetic elements. In addition the system contains two further genetic elements which are believed not to specify any protein products. These are the promoter and the operator, the functions of which we shall discuss below. The five elements are arranged along the DNA in the order:

$$\begin{matrix} \text{repressor} \\ \text{gene} \end{matrix} - \text{promoter} - \text{operator} - \begin{matrix} \beta\text{-galactosidase} \\ \text{gene} \end{matrix} - \begin{matrix} \text{permease} \\ \text{gene} \end{matrix}$$

and we can now begin to describe the working of the system.

In the absence of a galactoside inducer the repressor (which is the protein product of the repressor gene) binds to the operator DNA. This highly specific interaction between protein and DNA prevents transcription by the RNA polymerase, which can initiate synthesis of mRNA only by binding to the promoter site on the DNA. Hence in the absence of an inducer no specific mRNA is formed. Inducers act by becoming attached to the repressor. This complex of inducer and repressor can no longer bind to the operator; in these circumstances the RNA polymerase can bind to the promoter and synthesize a mRNA molecule that corresponds to both β-galactosidase and the permease. This mRNA is translated by the ribosomes, and thus in the presence of an inducer β-galactosidase and galactoside permease are synthesized.

We remarked earlier that this system must not be taken as a model of all control systems, even in bacteria. However it contains several features that are highly characteristic of protein synthesis in bacteria. You will see that the mRNA that is synthesized from the promoter corresponds to *both* β-galactosidase *and* galactoside permease. A mRNA of this sort, corresponding to more than one protein, is called a *polycistronic* mRNA. (A 'cistron' is more or less equivalent to a gene in the sense that we used that term on p. 185). Such polycistronic mRNA molecules are common in bacteria: a striking example is that for the histidine-biosynthetic enzymes, where the mRNA corresponds to ten enzymes. Plainly in systems involving either biosynthetic or degradative sequences it is extremely desirable for the synthesis of all the enzymes to be controlled together: either histidine is present in the medium, in which case none of the biosynthetic enzymes are needed, or it is absent, in which case all of them are needed together. It may be for this reason that the genes for a particular pathway of synthesis or degradation often lie together (p. 186). A set of genes transcribed into polycistronic mRNA is called an *operon*.

Another general point that we may derive from considering the 'lactose' system is that each gene or operon probably has its own promoter site at which the RNA polymerase binds and begins transcription. If these sites differ in affinity for the polymerase, one can readily see how it is that different genes or operons are expressed at different rates. For example, the repressor gene is expressed at a very low rate, since repressor needs to be made only in extremely small quantities (there are only about a dozen repressor molecules in each cell, as compared with about 10^5 molecules of β-galactosidase in a fully induced cell); this suggests that the repressor gene has a promoter with low affinity for the RNA polymerase.

With some operons, including the 'lactose' system, the structure of the promoter has an additional significance. We have already said that RNA polymerase will bind to the promoter of the 'lactose' operon only if a galactoside inducer is present to form a complex with the repressor. It turns out that there is an additional factor needed for binding of the RNA polymerase to the promoter: the compound 3′,5′-cyclic AMP (see Fig. 22.1). This compound is formed from ATP by the enzyme adenyl cyclase, and, as we shall see in the next chapter, it is of great importance in the regulation of enzyme *activity*, particularly in mediating the action of some hormones in higher animals. Here we are concerned with its role in enzyme *synthesis*, and we can summarize that role as follows. Under conditions where bacteria have a good source of energy and are able

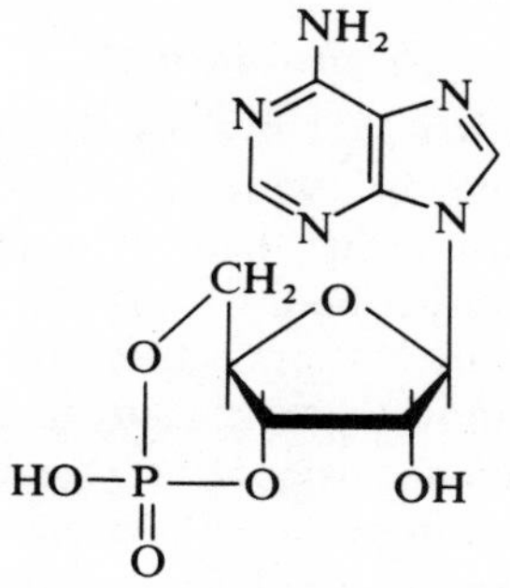

Fig. 22.1. 3′,5′-Cyclic AMP.

to grow rapidly (for example when glucose is present), they appear to maintain a low intracellular concentration of 3′,5′-cyclic AMP. When glucose is available it is to the advantage of the cells not to synthesize enzymes, such as β-galactosidase, whose function is to break down alternative carbon sources. Such enzymes are encoded by operons (which we might call 'catabolic' operons) that seem always to require 3′,5′-cyclic AMP for their transcription. In other words the promoters of these operons allow RNA polymerase to bind only if 3′,5′-cyclic AMP is present in sufficiently high concentration – and this concentration, it seems, is achieved only when the cells are not growing so fast on a rich medium as to make the transcription of such 'catabolic' operons superfluous. The result is that – to take the 'lactose' operon as an example – frequent transcription by the RNA polymerase occurs only when a galactoside inducer is *present, and* when glucose is absent.

There is a further feature of the 'lactose' system which we have not hitherto mentioned but which seems to be of general applicability in bacteria. When a galactoside inducer is added to a culture of bacteria the synthesis of β-galactosidase and the permease begins very quickly; equally when the inducer is removed synthesis ceases very quickly. The rapid cessation seems to be due to rapid breakdown of mRNA, and in general it is found that in bacteria most mRNA is degraded very rapidly. This is in part the reason why bacteria are able to adapt quickly to changes in the medium by changing their pattern of protein synthesis: once a particular species of mRNA is degraded the ribosomes that were translating it will be free to translate newly synthesized mRNA (see p. 204). Thus in order to maintain induction of a particular set of genes, the corresponding mRNA must be synthesized continuously.

Higher animals

We began our discussion of regulation in bacteria by considering what would happen if bacteria were placed in media with or without histidine, or with lactose rather than glucose as carbon source. The systems that we have described, with rapid changes in the rate of synthesis of different mRNA molecules and rapid degradation of pre-existing mRNA, allow bacteria easily to adapt to changes in the medium. In higher animals, by contrast, the ability to adapt rapidly to changes in the cell's environment is comparatively unimportant. It is true that changes in diet may occur, and these do in fact result in modest increases or decreases in the rate of synthesis of some enzymes in the liver – the liver being the organ that must deal directly with the products of digestion of the food. But most of the cells of a higher animal are continuously bathed in a fluid whose composition is kept more or less constant; there is no question of histidine, or any other amino acid, suddenly appearing or disappearing.

Consequently the cells of higher animals do not, for the most part, possess mechanisms for rapidly switching on or off the expression of their genes. Nonetheless there *are* mechanisms for controlling gene expression that are of great importance in animal cells; these are concerned not with the ability to respond to changes in the environment but rather with permanent differentiation into various cell types.

In Chapter 21 we gave a description of a hepatocyte, which we chose to regard as a 'typical' animal cell precisely because its functions are comparatively unspecialized. This lack of specialization at the functional level is reflected in its complement of enzymes: a hepatocyte makes a large number of different enzymes, or, to put it another way, a large number of its genes are being expressed. Many other cells of the animal body are more specialized than the hepatocyte, and we referred in Chapter 4 to some of their products. For instance, a muscle cell makes large quantities of two proteins, actin and myosin, which are necessary for contractility; fibroblasts (in connective tissue) make large quantities of the protein, collagen, which is formed into fibres; plasma cells (in the lymphatic system) make large quantities of immunoglobulins (see p. 36). Other specialized cells make particular proteins that have non-enzymic functions: reticulocytes (the precursors of red blood cells) make haemoglobin, and the cells of endocrine glands make the specific hormones that are characteristic of those glands (such as insulin, synthesized by the β cells of the islets of Langerhans in the pancreas).

In all these cases, a particular gene or set of genes is expressed in one cell type at a relatively high rate and is silent in other cell types in the animal. Thus differentiation of cell types, which is often thought of in terms of the microscopic appearance of the cells, can be regarded at the molecular level as resulting from certain genes being either silenced or activated.

In discussing the mechanism of differentiation, the first question we must consider is whether, in a cell that is not expressing a particular gene, that gene is in fact still present. Does differentiation result from the *loss* of portions of genetic material that are of no use to the specialized cell? There is now abundant evidence that many, if not all, differentiated cells retain the total genetic complement of the fertilized egg that gave rise to all the tissues of the body. Consequently, the fact that in any given differentiated cell most of the genes are not expressed – that is, that fibroblasts (for example) do not make haemoglobin, hormones, digestive enzymes or immunoglobulins – must be due to the permanent repression of these genes rather than to their absence.

About the means by which this repression is brought about we know very little. One might be tempted to think that the mechanism of repression of animal genes is similar to that in bacteria which we discussed above, but such a temptation should be avoided. In the first place, as we have already stressed, the biological significance of differentiation of cell types is quite different from that of induction and repression in bacteria. Secondly, the organization of the genetic material in higher organisms is far more complicated than in bacteria (p. 61 and p. 187). Thirdly, the transcription of animal genes does not give rise directly to mRNA, as it does with bacterial genes; rather, the primary transcript is extensively processed by special enzymes before becoming the mRNA that is translated at the ribosomes. We have good reason to believe that, where the protein product of a particular gene is not synthesized by a given tissue, the corresponding mRNA is often absent too. But we do not know whether the gene is not transcribed at all, or whether, although it is transcribed, the transcript fails to become mRNA. There is also evidence that the repression of a set of genes is sometimes dependent on substances passed from the cytoplasm into the nucleus, but (except in a few cases – see p. 239) we do not know what these substances are or how they came to be in the cytoplasm in the first place. On the intriguing subject of differentiation it is, as yet, far easier to ask questions than to provide answers.

23 Control of enzymic activity, and an introduction to hormones

In the chapters on intermediary metabolism, we described pathways for the degradation and synthesis of several compounds. We outlined, for example, the route of degradation of glycogen, and also the route of synthesis (which, as we remarked, is slightly different). However, any given cell, even though it contains the enzymes of both pathways, will not be rapidly degrading and synthesizing glycogen at the same moment. There must therefore be some control system that enables one pathway to be active, and the other inhibited, according to requirements.

In a similar way there must be means of preventing the synthesis of intermediates or building-blocks that are not required. To refer again to histidine synthesis in bacteria, we have already seen that the addition of histidine to the medium represses the formation of the enzymes required for histidine synthesis. However, the enzymes that are already present in the cell will be expected to be active for a comparatively long time, and will, unless there is a rapid means of rendering them inactive, waste both raw materials and energy in making a superfluous compound. In fact is is found that if an amino acid, purine or pyrimidine is added to a culture of bacteria it will generally inhibit its own synthesis. This inhibition occurs so rapidly that repression of enzyme synthesis is insufficient to account for the observed result.

These examples suggest that there are means of inhibiting or enhancing enzymic *activity*, in addition to the means of repressing and inducing enzyme *synthesis* that we described in the last chapter. The best-studied examples are those that we mentioned in the preceding paragraph, namely the inhibition by amino acids, purines or pyrimidines of their own synthesis (often called end-product inhibition), and we shall say something about this first. Later we shall apply some of these principles in describing the way in which the degradation and synthesis of carbohydrates are controlled.

End-product inhibition differs in two important respects from repression. First, the end-product inhibits not all the enzymes leading to its synthesis but only the first enzyme that is specific to the pathway. Secondly, this inhibition is a property of the affected enzyme itself, so that it can be observed outside the cell in a purified enzyme preparation. A detailed discussion of end-product inhibition would require an account of enzyme kinetics, which is beyond the scope of this book; but we can outline some general principles even without considering in depth the behaviour of the inhibited enzyme.

One of the best characterized of the enzymes that are subject to end-product inhibition is aspartate transcarbamylase, which is the first enzyme in the pathway of pyrimidine synthesis (see p. 192). The $\Delta G^{0\prime}$ of the reaction catalysed by this enzyme overwhelmingly favours the synthesis of carbamyl aspartic acid. This is an important feature of reactions that are subject to end-product inhibition. We have seen throughout this book that 'reversible' reactions (that is those in which the overall energy balance allows the formation of a significant amount of products in either direction) are generally used both in degradative and in synthetic pathways. Therefore to subject an enzyme that catalyses one of these reactions to end-product inhibition, or indeed to any other inhibition or activation, would have no effect in controlling the rate of degradation or synthesis as a whole. We shall refer to this point again in discussing the control of degradation and synthesis of carbohydrates.

The enzyme aspartate transcarbamylase is inhibited by CTP, which is one of the end-products of the pathway of pyrimidine synthesis. The effect of CTP on the enzyme is an interesting one. The first stage in an enzyme reaction is binding of the substrate (see p. 47); and this enzyme has to bind both aspartic acid and carbamyl phosphate. It is found that in the presence of CTP the enzyme undergoes a change in its quaternary structure, and the result of this is that its affinity for aspartic acid is much diminished. Thus at a given concentration of aspartic acid the rate of reaction is much less in the presence of CTP than in its absence. (The change in the configuration of aspartate transcarbamylase is reversible, so that removal of CTP will restore its activity to normal.) The molecular basis for this phenomenon is analogous to the mechanism by which the binding of one molecule of oxygen to haemoglobin alters its affinity for oxygen molecules that arrive subsequently (see p. 42).

This kind of enzymic property is obviously useful in an organism that may suffer changes in its environment. For example a bacterium will, in the absence of pyrimidines from the medium, make its own pyrimidines by the pathway that we outlined on pp.

192–4, which includes the reaction catalysed by aspartate trans-carbamylase; if a pyrimidine is supplied in the medium some of it will be converted to CTP and this will inhibit the enzyme. In addition end-product inhibition is an extremely valuable control for internal regulation of a cell's metabolism. Suppose, for example, that the rate of RNA synthesis in a cell synthesizing its own pyrimidines is, for some reason, reduced. This reduction will result in an accumulation of the precursors of RNA synthesis, one of which is CTP. End-product inhibition will now diminish the rate of pyrimidine synthesis until the concentration of CTP falls to its normal level, when pyrimidine synthesis will be quickly resumed.

Changes in affinity of enzymes for their substrates are a common means of regulating metabolic activity. They are frequently brought about by end-products (as with aspartate transcarbamylase) and then are generally inhibitory – i.e. the end-product diminishes the affinity of an enzyme for its substrate. Sometimes, however, they are brought about by other substances, and they may then involve either a decrease or an increase in affinity. Interesting examples come from studies of the control of glycolysis and of carbohydrate synthesis.

Glycolysis is a process that requires control for reasons of the cell's internal economy. There is no end-product which, like a pyrimidine, may suddenly turn out to be in large excess in the medium that bathes the cell. On the other hand, glycolysis is crucial for supplying energy and thus in regulating, in a very general way, all the cell's activities. Its control is obviously essential in metabolism.

Now although glycolysis has no true end-product, there is a sense in which ATP can be regarded as its end-product. In a situation where ATP is abundant the need for glycolytic activity is obviously greatly reduced. Conversely when ATP is in short supply, glycolysis requires to be stimulated.

If we compare the glycolytic pathway (Chapter 10) with the pathway for synthesizing carbohydrate (Chapter 15), we shall see that most of the reactions occur in both – that is to say that they can achieve a net formation of products in either direction. There is, however, one reaction, placed at a crucial stage, that has a high negative $\Delta G^{0\prime}$, namely the phosphorylation of fructose-6-phosphate by ATP, catalysed by phosphofructokinase. Whether the substrate of glycolysis is glucose or glycogen or galactose, phosphofructokinase provides a potential point of control for the whole process. It is therefore not surprising that phosphofructokinase is inhibited by ATP. Moreover, the three adenine nucleotides in the cell are in equilibrium owing to a reaction catalysed by the enzyme adenylate kinase:

$$ATP + AMP \rightleftharpoons 2ADP$$

so that when the concentration of ATP is high the concentration of AMP is low. Phosphofructokinase is found to be greatly stimulated by AMP as well as inhibited by ATP. Consequently when the concentration of ATP is low – that is when the conditions demand an increase in the rate of glycolysis – the concentration of AMP rises, so that phosphofructokinase is stimulated to convert more fructose-6-phosphate to fructose-1,6–bisphosphate.

Exactly the converse arguments apply to the synthesis of glycogen from lactic acid or from intermediates in the Krebs cycle. Under conditions where the supply of ATP is plentiful, and therefore (see the reaction above) the concentration of AMP is low, it is desirable for the cell to be able to synthesize carbohydrate from simpler precursors and thus build up its stores of substrate. When the concentration of ATP falls, and that of AMP rises, it is desirable to inhibit the synthesis of carbohydrate. Now we have seen (p. 163) that one of the reactions that is specific to carbohydrate synthesis is the hydrolysis of fructose-1,6-bisphosphate to fructose-6-phosphate catalysed by fructose bisphosphatase; and it is found that this enzyme is strongly inhibited by AMP.

Hormones

The kind of mechanism that we have described has considerable flexibility, and might well be adequate to meet the needs of a single-celled organism in respect of regulation of enzyme activity. But in a complex organism with millions of cells, even these mechanisms would probably be insufficient to ensure an integrated response in several organs. For instance, the characteristic changes that occur when an animal reaches sexual maturity require that many different tissues alter their biochemical behaviour simultaneously. Or, to give a quite different example, when a mammal begins vigorous exercise after a period of rest, both the heart and many separate blocks of skeletal muscles need to increase their metabolic activity within a few seconds. Co-ordination of these changes in various tissues at the same time is due in large part to the action of hormones.

Hormones are quite varied both in their chemical structures and in their biochemical effects, but it is useful to divide them into two classes: those that enter the cells of the target organs, and those that bind to the cell surface.

Many of the hormones that enter the cells on which they exert their effect are *steroids* (p. 85), which, being lipid soluble, can pass through cell membranes. Perhaps the best-understood group is the female sex hormones. Inside the cells of their target organs – such as the uterus and the vagina – these steroids bind to specific protein receptors in the soluble cytoplasm. (Cells of organs that are not sensitive to the hormones do not possess these specific receptors.) The hormone–receptor complex then enters the nucleus, where it binds to chromatin. Although we are not certain of the details of the mechanism of action of this complex, it is known that certain proteins are synthesized in large quantities in response to the hormones, and we may infer that particular genes are activated in some way by the binding of the hormone–receptor complex to chromatin.

Many hormones, on the other hand, such as insulin, glucagon and adrenalin (sometimes called epinephrine), also bind to specific receptors, but in these cases the receptors are located on the cell surface. The binding of the hormone to its receptor provokes changes within the target cell, and these changes result in the rearrangement of metabolic activity that is characteristic of the effect of the hormone.

We can illustrate this kind of mechanism by describing the action of adrenalin, which has been extensively studied in recent years and which is probably the best understood of all hormones. It has been known for some time that adrenalin stimulates the breakdown of glycogen in muscle and liver. The enzyme phosphorylase (p. 131) which is responsible for this breakdown exists in two forms, form *a* which is active and form *b* which is much less active. We now know that the binding of adrenalin to its receptor on the cell membrane activates an enzyme called adenylate cyclase, which is bound to the inner surface of the membrane. Adenylate cyclase catalyses the synthesis of 3′,5′-cyclic AMP (Fig. 22.1) from ATP: 3′,5′-cyclic AMP activates an enzyme called protein kinase, which in turn activates another enzyme called phosphorylase *b* kinase, which finally converts phosphorylase *b* into phosphorylase *a* (see Fig. 23.1). We may ask why so complex a system is needed to provide a mechanism of action of a hormone. The answer is that each stage of this system provides an *amplification* of the effect. Only a few molecules of adrenalin are needed to activate adenylate cyclase, which produces many more molecules of 3′,5′-cyclic AMP to

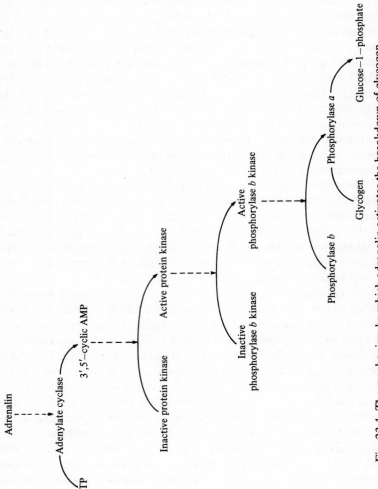

Fig. 23.1. The mechanism by which adrenalin activates the breakdown of glycogen.

activate protein kinase, which in turn can activate still more molecules of phosphorylase *b* kinase, and so on.

3',5'-Cyclic AMP is now known to be involved in the mechanism of action of many hormones, including parathyroid hormone, glucagon, vasopressin and a group of several hormones synthesized by the anterior pituitary. In each case the hormone binds to specific receptors on the surface of those cells that are responsive to it. This binding always has the effect of provoking the formation of 3',5'-cyclic AMP by adenylate cyclase, but the way in which 3',5'-cyclic AMP will then act depends on the complement of enzymes in the target cell, and in most cases we do not understand the sequence of biochemical events nearly so well as we do the breakdown of glycogen that is stimulated by adrenalin.

Insulin has been known for more than fifty years to be the hormone that is either missing or not effective in patients suffering from diabetes mellitus, a disease in which the metabolism of glucose is disordered. Diabetics characteristically have in their blood an excessively high concentration of glucose, and the administration of insulin causes this concentration to fall promptly. Insulin has many other effects as well. It promotes the synthesis of protein from amino acids and of triglycerides and glycogen from glucose, and several other changes in the pattern of metabolism have been ascribed to it. One explanation that has been proposed for many of these effects is to suggest that insulin stimulates the entry of glucose and other substances into its target cells, which are known to have specific insulin receptors. However, even though insulin probably does have some action on membrane transport, it is not known whether this is its primary effect nor whether such an action accounts for all its metabolic activities.

Conclusion

The reader may have noticed the style of this book becoming more tentative in the last two or three chapters, compared with the more confident tone of (for example) Section II. We have been discussing areas of biochemistry in which the systems are highly complex and interrelated, and the details of these are as yet quite unclear. The studies of differentiation and of hormone action are among the most difficult that biochemists have yet tackled, and there is a long way to go before we understand these subjects at the molecular level. Even so, they have begun to yield their secrets, and one may hope that at least their principles will be clearly formulated within the next decade.

Index

acetaldehyde, 130
acetoacetic acid, 143
acetoacetyl coenzyme A, 143
acetoacetyl group, in synthesis of fatty
    acids, 168
acetyl coenzyme A, high-energy
    compound, 103
    a central compound in intermediary
    metabolism, 134
    from ketogenic amino acids, 177–8
    in Krebs cycle, 135, 142
    not convertible to Krebs cycle inter-
    mediates, 165
    from oxidation of fatty acids, 142
    from pyruvic acid, 133–4
    in synthesis of fatty acids, 167, 168
acetyl coenzyme A carboxylase, 167
acetyl (two-carbon) fragments
    from breakdown of polysaccharides
    and fats, 121, 122, and from oxida-
    tion of pyruvic acid, 133
    in intermediary metabolism, 124
    in synthesis of fatty acids, 122, 123
acetyl galactosamine, 64
acetyl glucosamine, 64
acetyl muramic acid, 65
acetyl neuraminic acid, 65
acids (two- to six-carbon), in inter-
    mediary metabolism, 121, 122,
    123, 124
aconitase, 135
actin, muscle protein, 34, 234
activation reactions, effectively irrever-
    sible, 141
    for fatty acid breakdown, 140–1
    for polysaccharide synthesis, 164–5
    for protein synthesis, 178
acyl carrier proteins, 168, 169
acyl mercaptide high-energy bonds, 136
acyl phosphates, high-energy com-
    pounds, 103, 128
adenine, 52, 189
    hydrogen-bonding of, to thymine or
    uracil, 53
    isopentenyl and threonyl derivatives
    of, 59

adenine nucleotides, equilibrium
    between, 240
adenosine diphosphate (ADP), high-
    energy compound, 101, 103
    antiport of ATP and, across mito-
    chondrial membrane, 223
    phosphorylation of, to ATP, see
    phosphorylation
adenosine monophosphate (AMP), 101,
    140, 178
    inhibits fructose bisphosphatase, 240
    stimulates phosphofructokinase, 240
    synthesis of, 190–1
adenosine 3′,5′-cyclic monophosphate
    (cAMP)
    activates protein kinase, 241, 242
    involved in action of many hormones,
    241, 242, 243
    required for binding of RNA poly-
    merase to DNA, 232–3
adenosine triphosphatase (ATPase)
    in mitochondrial inner membrane,
    118
    subunit of myosin as, 34
adenosine triphosphate (ATP), high-
    energy compound, 100–1, 103,
    major energy carrier of cell, 118
    antiport of ADP and, across mito-
    chondrial membrane, 223
    inhibits phosphofructokinase, 239
    in muscle, 34
    phosphorylation of ADP to, see
    phosphorylation
    a precursor of RNA, 189
    produced: in glycolysis, 128, 129,
    130; in Krebs cycle, 138, 177; in
    mitochondria, 219; in oxidations,
    107, 112–14, 142; in photosyn-
    thesis (light reaction), 151, 152,
    153, 154
    required: in activation of amino acids,
    178, and of fatty acids, 140, 142–3;
    in carboxylation of pyruvic acid,
    139, 161; to drive transport against
    concentration gradient, 224; in
    fixation of ammonia in carbamyl

245